NATURAL HISTORY
UNIVERSAL LIBRARY

西方博物学大系

主编：江晓原

A HISTORY OF THE BRITISH STALK-EYED CRUSTACEA

不列颠甲壳亚门动物志

[英] 托马斯·贝尔 著

华东师范大学出版社

图书在版编目（CIP）数据

不列颠甲壳亚门动物志 = A history of the British Stalk-eyed Crustacea：英文 /（英）托马斯·贝尔（Thomas Bell）著. — 上海：华东师范大学出版社, 2018
（寰宇文献）
ISBN 978-7-5675-7991-0

Ⅰ.①不… Ⅱ.①托… Ⅲ.①甲壳纲-动物志-英文 Ⅳ.①Q959.223

中国版本图书馆CIP数据核字(2018)第154461号

不列颠甲壳亚门动物志

A history of the British Stalk-eyed Crustacea
（英）托马斯·贝尔（Thomas Bell）

特约策划	黄曙辉　徐　辰
责任编辑	庞　坚
特约编辑	许　倩
装帧设计	刘怡霖

出版发行	华东师范大学出版社
社　　址	上海市中山北路3663号　邮编 200062
网　　址	www.ecnupress.com.cn
电　　话	021-60821666　行政传真 021-62572105
客服电话	021-62865537
门市（邮购）电话	021-62869887
地　　址	上海市中山北路3663号华东师范大学校内先锋路口
网　　店	http://hdsdcbs.tmall.com/

印 刷 者	虎彩印艺股份有限公司
开　　本	787×1092　16开
印　　张	29.5
版　　次	2018年8月第1版
印　　次	2018年8月第1次
书　　号	ISBN 978-7-5675-7991-0
定　　价	498.00元（精装全一册）

出 版 人　王　焰

（如发现本版图书有印订质量问题，请寄回本社客服中心调换或电话021-62865537联系）

《西方博物学大系》总序

江晓原

《西方博物学大系》收录博物学著作超过一百种，时间跨度为15世纪至1919年，作者分布于16个国家，写作语种有英语、法语、拉丁语、德语、弗莱芒语等，涉及对象包括植物、昆虫、软体动物、两栖动物、爬行动物、哺乳动物、鸟类和人类等，西方博物学史上的经典著作大备于此编。

中西方"博物"传统及观念之异同

今天中文里的"博物学"一词，学者们认为对应的英语词汇是Natural History，考其本义，在中国传统文化中并无现成对应词汇。在中国传统文化中原有"博物"一词，与"自然史"当然并不精确相同，甚至还有着相当大的区别，但是在"搜集自然界的物品"这种最原始的意义上，两者确实也大有相通之处，故以"博物学"对译Natural History一词，大体仍属可取，而且已被广泛接受。

已故科学史前辈刘祖慰教授尝言：古代中国人处理知识，如开中药铺，有数十上百小抽屉，将百药分门别类放入其中，即心安矣。刘教授言此，其辞若有憾焉——认为中国人不致力于寻求世界"所以然之理"，故不如西方之分析传统优越。然而古代中国人这种处理知识的风格，正与西方的博物学相通。

与此相对，西方的分析传统致力于探求各种现象和物体之间的相互关系，试图以此解释宇宙运行的原因。自古希腊开始，西方哲人即孜孜不倦建构各种几何模型，欲用以说明宇宙如何运行，其中最典型的代表，即为托勒密（Ptolemy）的宇宙体系。

比较两者，差别即在于：古代中国人主要关心外部世界"如何"运行，而以希腊为源头的西方知识传统（西方并非没有别的知识传统，只是未能光大而已）更关心世界"为何"如此运行。在线性发展无限进步的科学主义观念体系中，我们习惯于认为"为何"是在解决了"如何"之后的更高境界，故西方的分析传统比中国的传统更高明。

然而考之古代实际情形，如此简单的优劣结论未必能够成立。例如以天文学言之，古代东西方世界天文学的终极问题是共同的：给定任意地点和时刻，计算出太阳、月亮和五大行星（七政）的位置。古代中国人虽不致力于建立几何模型去解释七政"为何"如此运行，但他们用抽象的周期叠加（古代巴比伦也使用类似方法），同样能在足够高的精度上计算并预报任意给定地点和时刻的七政位置。而通过持续观察天象变化以统计、收集各种天象周期，同样可视之为富有博物学色彩的活动。

还有一点需要注意：虽然我们已经接受了用"博物学"来对译 Natural History，但中国的博物传统，确实和西方的博物学有一个重大差别——即中国的博物传统是可以容纳怪力乱神的，而西方的博物学基本上没有怪力乱神的位置。

古代中国人的博物传统不限于"多识于鸟兽草木之名"。体现此种传统的典型著作，首推晋代张华《博物志》一书。书名"博物"，其义尽显。此书从内容到分类，无不充分体现它作为中国博物传统的代表资格。

《博物志》中内容，大致可分为五类：一、山川地理知识；二、奇禽异兽描述；三、古代神话材料；四、历史人物传说；五、神仙方伎故事。这五大类，完全符合中国文化中的博物传统，深合中国古代博物传统之旨。第一类，其中涉及宇宙学说，甚至还有"地动"思想，故为科学史家所重视。第二类，其中甚至出现了中国古代长期流传的"守宫砂"传说的早期文献：相传守宫砂点在处女胳膊上，永不褪色，只有性交之后才会自动消失。第三类，古代神话传说，其中甚至包括可猜想为现代"连体人"的记载。第四类，各种著名历史人物，比如三位著名刺客的传说，此三名刺客及所刺对象，历史上皆实有其人。第五类，包括各种古代方术传说，比如中国古代房中养生学说，房中术史上的传说人物之一"青牛道士封君达"等等。前两类与西方的博物学较为接近，但每一类都会带怪力乱神色彩。

"所有的科学不是物理学就是集邮"

在许多人心目中，画画花草图案，做做昆虫标本，拍拍植物照片，这类博物学活动，和精密的数理科学，比如天文学、物理学等等，那是无法同日而语的。博物学显得那么的初级、简单，甚至幼稚。这种观念，实际上是将"数理程度"作为唯一的标尺，用来衡量一切知识。但凡能够使用数学工具来描述的，或能够进行物理实验的，那就是"硬"科学。使用的数学工具越高深越复杂，似乎就越"硬"；物理实验设备越庞大，花费的金钱越多，似乎就越"高端"、越"先进"……

这样的观念，当然带着浓厚的"物理学沙文主义"色彩，在很多情况下是不正确的。而实际上，即使我们暂且同意上述"物理学沙文主义"的观念，博物学的"科学地位"也仍然可以保住。作为一个学天体物理专业出身，因而经常徜徉在"物理学沙文主义"幻影之下的人，我很乐意指出这样一个事实：现代天文学家们的研究工作中，仍然有绘制星图，编制星表，以及为此进行的巡天观测等等活动，这些活动和博物学家"寻花问柳"，绘制植物或昆虫图谱，本质上是完全一致的。

这里我们不妨重温物理学家卢瑟福（Ernest Rutherford）的金句："所有的科学不是物理学就是集邮（All science is either physics or stamp collecting）。"卢瑟福的这个金句堪称"物理学沙文主义"的极致，连天文学也没被他放在眼里。不过，按照中国传统的"博物"理念，集邮毫无疑问应该是博物学的一部分——尽管古代并没有邮票。卢瑟福的金句也可以从另一个角度来解读：既然在卢瑟福眼里天文学和博物学都只是"集邮"，那岂不就可以将博物学和天文学相提并论了？

如果我们摆脱了科学主义的语境，则西方模式的优越性将进一步被消解。例如，按照霍金（Stephen Hawking）在《大设计》（*The Grand Design*）中的意见，他所认同的是一种"依赖模型的实在论（model-dependent realism）"，即"不存在与图像或理论无关的实在性概念（There is no picture- or theory-independent concept of reality）"。在这样的认识中，我们以前所坚信的外部世界的客观性，已经不复存在。既然几何模型只不过是对外部世界图像的人为建构，则古代中国人干脆放弃这种建构直奔应用（毕竟在实际应用中我们只需要知道七政"如何"运行），又有何不可？

传说中的"神农尝百草"故事，也可以在类似意义下得到新的解读："尝百草"当然是富有博物学色彩的活动，神农通过这一活动，得知哪些草能够治病，哪些不能，然而在这个传说中，神农显然没有致力于解释"为何"某些草能够治病而另一些则不能，更不会去建立"模型"以说明之。

"帝国科学"的原罪

今日学者有倡言"博物学复兴"者，用意可有多种，诸如缓解压力、亲近自然、保护环境、绿色生活、可持续发展、科学主义解毒剂等等，皆属美善。编印《西方博物学大系》也是意欲为"博物学复兴"添一助力。

然而，对于这些博物学著作，有一点似乎从未见学者指出过，而鄙意以为，当我们披阅把玩欣赏这些著作时，意识到这一点是必须的。

这百余种著作的时间跨度为15世纪至1919年，注意这个时间跨度，正是西方列强"帝国科学"大行其道的时代。遥想当年，帝国的科学家们乘上帝国的军舰——达尔文在皇家海军"小猎犬号"上就是这样的场景之一，前往那些已经成为帝国的殖民地或还未成为殖民地的"未开化"的遥远地方，通常都是踌躇满志、充满优越感的。

作为一个典型的例子，英国学者法拉在（Patricia Fara）《性、植物学与帝国：林奈与班克斯》（*Sex, Botany and Empire, The Story of Carl Linnaeus and Joseph Banks*）一书中讲述了英国植物学家班克斯（Joseph Banks）的故事。1768年8月15日，班克斯告别未婚妻，登上了澳大利亚军舰"奋进号"。此次"奋进号"的远航是受英国海军部和皇家学会资助，目的是前往南太平洋的塔希提岛（Tahiti，法属海外自治领，另一个常见的译名是"大溪地"）观测一次比较罕见的金星凌日。舰长库克（James Cook）是西方殖民史上最著名的舰长之一，多次远航探险，开拓海外殖民地。他还被认为是澳大利亚和夏威夷群岛的"发现"者，如今以他命名的群岛、海峡、山峰等不胜枚举。

当"奋进号"停靠塔希提岛时，班克斯一下就被当地美丽的土著女性迷昏了，他在她们的温柔乡里纵情狂欢，连库克舰长都看不下去了，"道德愤怒情绪偷偷溜进了他的日志当中，他发现自己根本不可能不去批评所见到的滥交行为"，而班克斯纵欲到了"连嫖妓都毫无激情"的地步——这是别人讽刺班克斯的说法，因为对于那时常年航行于茫茫大海上的男性来说，上岸嫖妓通常是一项能够唤起"激情"的活动。

而在"帝国科学"的宏大叙事中，科学家的私德是无关紧要的，人们关注的是科学家做出的科学发现。所以，尽管一面是班克斯在塔希提岛纵欲滥交，一面是他留在故乡的未婚妻正泪眼婆娑地"为远去的心上人绣织背心"，这样典型的"渣男"行径要是放在今天，非被互联网上的口水淹死不可，但是"班克斯很快从他们的分离之苦中走了出来，在外近三年，他活得倒十分滋润"。

法拉不无讽刺地指出了"帝国科学"的实质："班克斯接管了当地的女性和植物，而库克则保护了大英帝国在太平洋上的殖民地。"甚至对班克斯的植物学本身也调侃了一番："即使是植物学方面的科学术语也充满了性指涉。……这个体系主要依靠花朵之中雌雄生殖器官的数量来进行分类。"据说"要保护年轻妇女不受植物学教育的浸染，他们严令禁止各种各样的植物采集探险活动。"这简直就是将植物学看成一种"涉黄"的淫秽色情活动了。

在意识形态强烈影响着我们学术话语的时代，上面的故事通常是这样被描述的：库克舰长的"奋进号"军舰对殖民地和尚未成为殖民地的那些地方的所谓"访问"，其实是殖民者耀武扬威的侵略，搭载着达尔文的"小猎犬号"军舰也是同样行径；班克斯和当地女性的纵欲狂欢，当然是殖民者对土著妇女令人发指的蹂躏；即使是他采集当地植物标本的"科学考察"，也可以视为殖民者"窃取当地经济情报"的罪恶行为。

后来改革开放，上面那种意识形态话语被抛弃了，但似乎又走向了另一个极端，完全忘记或有意回避殖民者和帝国主义这个层面，只歌颂这些军舰上的科学家的伟大发现和成就，例如达尔文随着"小猎犬号"的航行，早已成为一曲祥和优美的科学颂歌。

其实达尔文也未能免俗，他在远航中也乐意与土著女性打打交道，当然他没有像班克斯那样滥情纵欲。在达尔文为"小猎犬号"远航写的《环球游记》中，我们读到："回程途中我们遇到一群黑人姑娘在聚会，……我们笑着看了很久，还给了她们一些钱，这着实令她们欣喜一番，拿着钱尖声大笑起来，很远还能听到那愉悦的笑声。"

有趣的是，在班克斯在塔希提岛纵欲六十多年后，达尔文随着"小猎犬号"也来到了塔希提岛，岛上的土著女性同样引起了达尔文的注意，在《环球游记》中他写道："我对这里妇女的外貌感到有些失望，然而她们却很爱美，把一朵白花或者红花戴在脑后的髻上……"接着他以居高临下的笔调描述了当地女性的几种发饰。

用今天的眼光来看，这些在别的民族土地上采集植物动物标本、测量地质水文数据等等的"科学考察"行为，有没有合法性问题？有没有侵犯主权的问题？这些行为得到当地人的同意了吗？当地人知道这些行为的性质和意义吗？他们有知情权吗？……这些问题，在今天的国际交往中，确实都是存在的。

也许有人会为这些帝国科学家辩解说：那时当地土著尚在未开化或半开化状态中，他们哪有"国家主权"的意识啊？他们也没有制止帝国科学家的考察活动啊？但是，这样的辩解是无法成立的。

姑不论当地土著当时究竟有没有试图制止帝国科学家的"科学考察"行为，现在早已不得而知，只要殖民者没有记录下来，我们通常就无法知道。况且殖民者有军舰有枪炮，土著就是想制止也无能为力。正如法拉所描述的："在几个塔希提人被杀之后，一套行之有效的易货贸易体制建立了起来。"

即使土著因为无知而没有制止帝国科学家的"科学考察"行为，这事也很像一个成年人闯进别人的家，难道因为那家只有不懂事的小孩子，闯入者就可以随便打探那家的隐私、拿走那家的东西、甚至将那家的房屋土地据为己有吗？事实上，很多情况下殖民者就是这样干的。所以，所谓的"帝国科学"，其实是有着原罪的。

如果沿用上述比喻，现在的局面是，家家户户都不会只有不懂事的孩子了，所以任何外来者要想进行"科学探索"，他也得和这家主人达成共识，得到这家主人的允许才能够进行。即使这种共识的达成依赖于利益的交换，至少也不能单方面强加于人。

博物学在今日中国

博物学在今日中国之复兴，北京大学刘华杰教授提倡之功殊不可没。自刘教授大力提倡之后，各界人士纷纷跟进，仿佛昔日蔡锷在云南起兵反袁之"滇黔首义，薄海同钦，一檄遥传，景从恐后"光景，这当然是和博物学本身特点密切相关的。

无论在西方还是在中国，无论在过去还是在当下，为何博物学在它繁荣时尚的阶段，就会应者云集？深究起来，恐怕和博物学本身的特点有关。博物学没有复杂的理论结构，它的专业训练也相对容易，至少没有天文学、物理学那样的数理"门槛"，所以和一些数理学科相比，博物学可以有更多的自学成才者。这次编印的《西方博物学大系》，卷帙浩繁，蔚为大观，同样说明了这一点。

最后，还有一点明显的差别必须在此处强调指出：用刘华杰教授喜欢的术语来说，《西方博物学大系》所收入的百余种著作，绝大部分属于"一阶"性质的工作，即直接对博物学作出了贡献的著作。事实上，这也是它们被收入《西方博物学大系》的主要理由之一。而在中国国内目前已经相当热的博物学时尚潮流中，绝大部分已经出版的书籍，不是属于"二阶"性质（比如介绍西方的博物学成就），就是文学性的吟风咏月野草闲花。

要寻找中国当代学者在博物学方面的"一阶"著作，如果有之，以笔者之孤陋寡闻，唯有刘华杰教授的《檀岛花事——夏威夷植物日记》三卷，可以当之。这是刘教授在夏威夷群岛实地考察当地植物的成果，不仅属于直接对博物学作出贡献之作，而且至少在形式上将昔日"帝国科学"的逻辑反其道而用之，岂不快哉！

2018 年 6 月 5 日
于上海交通大学
科学史与科学文化研究院

不列颠甲壳亚门动物志

出版说明

《不列颠甲壳亚门动物志》是英国学者托马斯·贝尔（Thomas Bell，1792—1880）的一部博物学著作。贝尔生于多塞特郡，自幼受其母熏陶，对博物学兴趣浓厚。二十岁时离乡学习，在伦敦成为一名牙医。1836年，他在教授解剖学课程的同时，为兼顾自己的爱好而受聘于国王学院，任动物学教授。1858年，荣任林奈学会会长。1836年底，当达尔文结束小猎犬号的远航考察返回伦敦后，贝尔立即与其接洽并接下后续科考报告工作中的爬行纲研究任务。1839年3月，他对加拉帕罗斯陆龟产地的结论有力推动了达尔文的演化论理论。此外，贝尔还协助整理编纂了小猎犬号远航报告中的动物学卷。同时，达尔文带回的甲壳纲动物样本也委托他管理研究，由此他成为甲壳纲动物研究的翘楚。退休后，贝尔闲居塞耳彭，又对曾在此地撰写出不朽名著的自然主义者吉尔伯特·怀特发生兴趣，甚至还在1877年为《塞耳彭自然史》编辑出版了一个新版本。三年后，他在塞耳彭去世。

贝尔毕生的集大成之作，就是这部《不列颠甲壳亚门动物志》。在本书中，他对不列颠岛出产的数百种甲壳亚门动物进行详尽分类介绍，每种均配以笔触老到精细的钢笔画，令科学研究与艺术之美浑然一体，美不胜收。今据原版影印。

BRITISH

STALK-EYED CRUSTACEA.

SIDNEY I. SMITH,
New Haven, Conn.

A HISTORY

OF THE

BRITISH

STALK-EYED CRUSTACEA.

BY

THOMAS BELL, Sec. R.S., F.G.S., F.Z.S.,

PRESIDENT OF THE LINNEAN SOCIETY;
MEMBER OF THE PHILOMATHIC AND NATURAL HISTORY SOCIETIES OF PARIS;
OF THE IMPERIAL ACADEMY CÆSAR. LEOPOLD. NATURÆ CURIOSORUM; OF THE
ACADEMY OF SCIENCES OF PHILADELPHIA; OF THE NATURAL HISTORY
SOCIETY OF NEW YORK; HONORARY MEMBER OF THE ROYAL
ZOOLOGICAL SOCIETY OF DUBLIN, ETC., ETC.
PROFESSOR OF ZOOLOGY IN KING'S COLLEGE, LONDON.

ILLUSTRATED BY 174 WOOD-ENGRAVINGS.

LONDON:
JOHN VAN VOORST, 1, PATERNOSTER ROW.

M.DCCC.LIII.

LONDON:
WOODFALL AND KINDER,
ANGEL COURT, SKINNER STREET.

TO

PROFESSOR RICHARD OWEN,

THE FAITHFUL AND UNCHANGED FRIEND OF MANY YEARS,

THIS LITTLE WORK IS INSCRIBED

BY

THE AUTHOR,

AS A HUMBLE TOKEN OF HIS LASTING

RESPECT AND AFFECTION.

PREFACE.

I HAVE little to say in this Preface, beyond the expression of my sincere regret for the delay which has occurred in the publication of the work. That delay has arisen from causes which it would not interest the public to be informed of, and which I have no wish to put forward for the sake of deprecating the displeasure or disappointment which it may have excited.

A much more agreeable task is that of acknowledging, which I do with feelings of sincere gratification and deep thankfulness, the extensive and valuable assistance which I have received from so many of my fellow-labourers in the field of Natural History. Their names are mentioned in connection with their contributions, in various parts of the work; and it would be invidious to particularise them here, lest, through inadvertence, any should be omitted. To one and all I beg to offer the tribute of my grateful thanks.

SELBORNE, HANTS.
July, 1853.

INTRODUCTION.

The structure of the Crustacea is so little known to the students of Natural History in this country, and there are so few works which give even the most superficial information on the subject, that it appears very desirable and even necessary to introduce the study of the British species, by a brief account of the general organization and physiology of this class of animals. Not only indeed is the subject itself one of great interest, but without some such introductory information it would not be possible to comprehend the descriptions of the different genera and species; for it will be found that in scarcely any other class of animals, is there a greater variety of form and structure, or more striking apparent anomalies in the modifications of the typical plan of organization, or in some cases greater difficulties in ascertaining the true homologies of the different elements, than in the present.

It is not, indeed, a very easy matter even to express, in a clear and definite phrase, the characters which, whilst belonging strictly to all the forms of Crustacea, shall distinctly exclude those of the approximate ones; for the variations which occur in every organ and function, in the different groups belonging to the crustacean type, are so considerable, as to render it almost impossible to include them all within one common and well-defined expression. The typical characters are so astonishingly modified, in some cases being totally changed, and in others absolutely lost, that the inexperienced student examining some aberrant form by the test of the known typical characters, might find it impossible to refer it to its true relations, without an investigation of the intermediate affinities, and an acquaintance with the laws which regulate their variations.

The separation of the true EPIZOA from the Crustacea has indeed, in some measure, facilitated the arrangement of the latter class, and enabled the zoologist to restrict within intelligible limits the characters which belong to the group.

I shall therefore, in the following sketch, consider the CRUSTACEA, the EPIZOA, and the CIRRIPEDES, as constituting three distinct types of form; with this restriction the Crustacea may be defined as articulated animals, having each segment of the external skeleton furnished with articulated appendages; they are all of them free or locomotive; the respiration is branchial, and they are, with very few exceptions, aquatic in their habits; the circulation is carried on by means of a complete vascular system, and is of a mixed character, the blood being received into an aortic heart, both from the branchiæ and from the system, and circulated in a mixed or partially decarbonized condition. The nervous system resembles, in its general principles, that of the Insects. It is ganglionic, longitudinal, and generally distinctly developed. The sexes are separate.

Such are the general characters by which the Crustacea proper may be distinguished, and which appear to be sufficiently defined, as far as our present knowledge extends. A further insight into the structure of each system of organs, as existing in the different orders and families of the class, will show how various and startling are some of their modifications.

The construction of the skeleton in this class of animals is for the most part very distinct from that of all others, although in some of the abnormal forms there is a remarkable deviation from the typical structure, and a corresponding approximation to that of other classes; as, for instance, in the segments of certain *Isopoda*, which resemble, in general character, some forms amongst the *Myriapoda*. In the greater number of them, and especially in the higher forms, the tegumentary skeleton is formed of a hard, solid, calcareous crust, the earthy portion of which consists of carbonate of lime, with a small portion of phosphate of the same earth. The colours by which the crust is, in many cases, very beautifully marked, depend upon a pigment which pervades different parts of the substance, and offers various hues, and sometimes curious and grotesque markings, in different species. The colouring matter, in these as in most other animals, is more intense on the upper than on the under surface, the latter being, in many, nearly pure white, whilst the former is deeply and brightly coloured. The earthy matter is deposited upon, and produced by, an organized vascular membrane or *corium*. In many of the smaller Crustacea, even amongst the higher forms, as in most of the *Palæmonidæ* or prawns, and other allied families, as well as in most of the lower groups, as the *Isopoda*, and others, the crust retains its semi-transparent, elastic, and flexible nature, resembling thin horn or parchment, the earthy matter being deposited in very small quantities. Although this difference is not wholly correlative with the groups in which it principally obtains,—as for instance, in the genus *Palæmon*, in which the crust of some species, as the common prawn, has scarcely any earthy matter, whilst in others, it is almost as solidly calcareous as in the lobster itself,—yet it would appear to bear a near relation to their habits; the presence of the calcareous substance hardening and solidifying the skeleton, and thus rendering it an efficient protection against the rocks and waves of the more exposed parts of the sea, being found in the greatest proportion in species exposed to such agents; whilst the others are either small, active creatures, swimming with great ease and constancy in more open and shallow situations, or creeping safely amongst fuci or under stones, and other protecting substances, or even attaching themselves to the surface of different species of fish.

The annulose character, typical of the great group to which it gives its designation, has, in a great number of the species composing this class, reached its maximum of development. The segments which surround the body are more complete, and more separately movable, whilst they possess a greater degree of individual solidity than in any others. They are also furnished with articulated appendages; each segment, whether remaining distinct or intimately united to others, bearing a single pair, in a more or less developed, or in a merely rudimentary condition. In numerous instances, from this intimate union or soldering together of two or more segments, the only indication of their theoretical separate existence is the presence of the normal number of these appendages; but with this aid it rarely happens, in the higher forms of Crustacea, that they cannot be proved to exist.

The true *normal* number of the segments, taking the whole class, appears to be twenty-one, of which, according to our present knowledge, seven must be considered as belonging to the head, and an equal number respectively to the thorax and the abdomen. Now, although it is true that there is not a single known species in which all these segments are found in a distinct and tangible condition—there being in all the forms, more or fewer of them so inseparably united together as to offer no other means by which to predicate their existence, than those already alluded to—yet, on the other hand, there is not one which may not be found distinctly formed in some or other of the species. The appendages, too, which have already been slightly mentioned, are no less subject to the most extraordinary variation both of form and office; many of them serving in one case the purposes of locomotion, in another the reception and preparation of the food, in another the attachment of the branchiæ, in another the support and protection of the eggs. When, therefore, we consider the almost endless diversity of form, under which the species composing this class of animals appear, the astonishing discrepancy which exists in the forms and relative proportions of the different regions of the body, and other parts of their organization, for the performance of offices and functions equally various, and see that all these diversities are produced only by modifications of a typical number of parts, we cannot but be struck by so remarkable and interesting an illustration of the great economical law, as it may be termed, that *the typical structure of any group being given, the different habits of its component species or minor groups are provided for, not by the creation of new organs or the destruction of others, but by the modification, in form, structure, or place, of organs typically belonging to the group.*

Of this law numerous examples will be exhibited in the course of this work, in the structural characters of every order and of every family; but for the sake of offering a single comprehensible illustration, the various modifications of the thoracic appendages may be selected. The typical structure of these may be considered as subserving the purposes of locomotion. This is the office which they fulfil, either wholly or in part, in all cases; and in some instances the whole of them are thus employed. In the *Isopoda*, for instance, the body consists very principally of the seven thoracic segments, and their appendages constitute seven pairs of true feet. In the *Amphipoda* the first or second pairs become modified in the male into strong holders by the greater development of the hand, and the movable character of the terminal articulation, and its applicability to a strong corresponding process from the penultimate articulation. In several of the *Læmodipoda* five pairs only of the thoracic appendages are developed into members, of which the first and second pairs constitute true hands or graspers, and the third and fourth are destined to a totally different office; forming *respiratory sacs*, to supply the place of the abdominal appendages in the *Isopoda*, the abdomen in the present instance being reduced to a mere rudiment. In the *Decapoda* there are only five pairs of true thoracic members, and these answer to the five posterior segments of the thorax; but the appendages to the segments anterior to these are rendered subservient to mastication, or to the preparation of the food, in the form of *footjaws* or *pedipalps*. I have only enumerated a few of the more conspicuous modifications of these organs, for the purpose of conveying at a glance some idea of the extraordinary aberrations from the typical structure which will meet us at every step, in the investigation of these animals, whose habits and requirements are so varied.

The composition of the segments in the Crustacea, although modified to a great extent in the different forms, is yet susceptible of being reduced to a perfect theoretical idea. Indeed, in many forms, the parts of which each segment is composed are distinctly appreciable by careful examination; and it is found that these parts consist in two arches, a superior and an inferior, each of which is formed of two middle and two lateral pieces. The superior central pair, *a a*, constitute the *tergum*, the lateral are called *epimera*, *b b*. Of the inferior arch, the two central pieces form the *sternum*, *c c*, and to the lateral, *d d*, the name of *episternum* has been applied. As we have already seen, in enumerating the segments themselves which compose the different regions of the body, that some or other of them are always found to be so intimately combined together that their distinction is lost, so in the present case also, some or other of the theoretical elements of the segments are either actually wanting, or certain of them are so intimately united that the normal number cannot be distinguished.

It is also necessary, in order to obtain a correct idea of the actual structure of the skeleton or supporting organs in the Crustacea, to consider those processes of crustaceous matter which, in the form of internal lamina, form the parietes of the cells and canals which are found in the interior of these animals, and many of which serve the office of bones, as the solid surfaces to which the muscles are attached. These have received the name of *apodema*. "They arise in all cases from the junction of two contiguous pieces of one segment, or from the union of two rings. They are produced by a duplicature of the tegumentary membrane, which dips more or less deeply amongst the internal organs, and which becomes encrusted with calcareous matter with the rest of the shell; they are consequently always formed of two layers, soldered, as it were, together." *

Of the various segments composing the three principal portions of the body, the head, the thorax, and the abdomen, some are found always to support similar, or rather identical, organs. Thus the first cephalic segment or ring invariably bears the peduncle of the eyes, and the second, or antennary, as constantly supports a pair of the antennæ. Of those which follow, there are the most extraordinary and unlooked-for modifications in the different groups; and no one who has only formed a theoretical notion of these parts could recognise in the simple piece of which the whole cephalic region is composed in the *Edriophthalma*, or in the carapace or shell of the brachyurous *Decapoda*, as in the common crab for instance, the mere combination of two or more of the cephalic segments which in other forms are found to be distinct. For a full and clear account of all these modifications, the reader is referred to the admirable work of Dr. Milne Edwards, so often quoted and referred to.

This author has, with great propriety, considered the genus *Squilla* as offering the form in which the different segments before enumerated are most distinctly exhibited; but even in this form there are some which are, as it were, soldered together, and the normal number is consequently not to be traced. The first cephalic segment, which, as before observed, is invariably destined to support the ocular peduncles, and is therefore termed the ophthalmic segment, is here quite distinct from the second, which is also very distinctly articulated with the third; the latter is, however, confounded with the next, and the following ones are only to be distinguished by dissection.† But the

* Edw. Hist. Nat. des Crust. i. p. 18. † Ib. p. 15.

last eleven are complete and perfectly distinct, and each of them, without exception, bears its appropriate pair of members.

Amongst the higher forms of Crustacea, it is in the Brachyura, where the nervous system is found in the most concentrated condition, that the condensation of the rings of which the body is composed, is carried to the greatest extent. It is indeed somewhat difficult, at first sight, to determine the homologies of the segments of which the carapace, as it is termed, is theoretically composed. This large enveloping buckler in fact covers the whole of the thorax, and even the abdomen itself is folded underneath it, so that the whole animal is hidden, when viewed from above, by this extraordinary development of two of the cephalic segments; and although in the Brachyura the first two segments, the ophthalmic and the antennary, are soldered to the carapace, yet, as we find that in some other forms these two are entirely distinct, it would appear that the carapace is essentially composed of the third and fourth rings, composing what Dr. Milne Edwards terms the antenno-maxillary segment.

This remarkable portion of the tegumentary system, covering, as it does, the whole of the viscera, is found to be more or less distinctly divided into *regions*, which are indicated by elevations, separated from each other by grooves; and to these regions have been given names derived from the different organs which are immediately covered by them. As reference is frequently made to these regions in generic and specific descriptions, I here give an illustration of them.*

---

* The gastric or stomachal region is marked r s; the branchial, r b; the hepatic, r h; the genital, r g; the cardiac r c; the intestinal, r i.

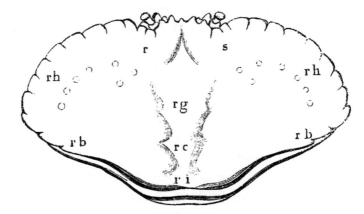

The thorax in the Decapods in general is externally only visible underneath, the upper part being covered by the carapace, and being in that part incomplete. The number of obvious segments in these higher forms is five, and as each segment bears its proper pair of appendages, which here are true ambulatory legs, the character of Decapods is thus produced. The superior surface of the thoracic segments is limited to the epimera, the tergum being absolutely wanting. Upon this upper surface on each side lie the branchiæ, or gills. In this brief sketch it is only necessary to refer to the apodemata as constituting the large cells of the thorax, formed by a dupli-

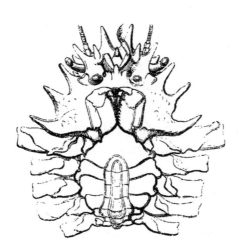

cature of the walls dipping into the thoracic cavity, and filled by the muscles which move the limbs.

The abdomen of the Brachyura is very moderately developed.* It folds entirely underneath the thorax, against which it is ordinarily closely applied. It consists, essentially, of seven segments, of which, however, in many cases, a greater or less number are so united as to be scarcely distinguishable. In the Macroura* they are far more extended, and serve the purposes of locomotion, being elongated, very moveable upon each other, and furnished at the extremity with a fan-shaped fin, formed of five pieces, of which the centre is the terminal abdominal segment.

In the lower forms, as the Edriophthalma, the rings of the body are more similar to each other, and constitute a nearly regular series of more or less perfect rings. Those of the head, however, are ordinarily much condensed, and soldered together; whilst the thorax consists of seven very distinct moveable segments, and the abdomen of either the same number, or nearly so; as in some cases the seventh is wanting, and in others the two anterior ones are united.

Between these two extreme cases, there are numerous intermediate modifications, which will be seen in the various families and genera.

The members or appendages to the different segments or annuli above described, form a very interesting and important part of the tegumentary system of these animals. Theoretically speaking, every segment has its pair of appendages, and, *vice versâ*, each pair of appendages,

* See the figures of the various species.

whenever they exist, presupposes a segment or ring to which they belong. In many cases, where a coalescence takes place between any of the contiguous segments, their distinct existence can only be predicated by the occurrence of the members which belong to them; thus, in the Brachyura, the carapace involves not only the third and fourth rings, enormously developed, but also the first two, which bear the eyes and antennæ, and which are indissolubly blended with the succeeding ones.

Normally there are twenty-one pairs of appendages or limbs: generally speaking, even in the higher forms, twenty only are perceived, as the terminal joint of the abdomen, which forms the central piece of the fan-like fin, has none which are perceptible. I have, however, observed them frequently in the common prawn, *Palæmon serratus*,* in the form of extremely minute points attached to the very extremity of the segment, and moveable.

The first pair exist only in the Podophthalma or stalk-eyed forms, and constitute the peduncles upon which the eyes are elevated; they are moveable, and in many cases are of considerable length, lying, when at rest, in grooves, or sockets, formed for their reception. The two following pairs are of great importance, forming, in most cases, organs of sense. These are the antennæ. One or both pairs exist in all the forms of true Crustacea; ordinarily

---

* I have often separated the whole twenty-one pairs of appendages in this species, and placed them *seriatim* on a card. They consist very clearly of the ocular peduncles, the anterior and posterior antennæ, the mandibles, the two pairs of maxillæ, the three pairs of foot-jaws, the five pairs of thoracic legs, the five pairs of abdominal false feet, the appendages to the sixth abdominal segment forming the lateral caudal flap, and the two minute rudimentary appendages above alluded to.

they are slender, elongated, moveable, and multiarticulate. They are, however, subject, in some forms even of the higher orders, to extraordinary modifications; thus in the genera *Scyllarus* and *Ibacus*, the external pair are developed into broad, flat organs of natation, and probably also constitute a pair of shovels for the purpose of burrowing: and in some Amphipoda, they are much elongated, serving as a pair of swimming or sustaining arms. The fourth pair always appertain to the mouth, and form manducating organs: these are the mandibles. The two pairs of jaws, or maxillæ, follow, and are also employed in the comminution of the food. Theoretically speaking, the next pair ought to be considered as belonging to the cephalic division of the body; these, as well as the previous and two following pairs, are, in the Decapoda, subservient to nutrition. The eighth and ninth pairs are, therefore, properly speaking, the first and second thoracic members, and, with the seventh, constitute the three pairs of footjaws or pedipalps, leaving, in this particular class, the five remaining thoracic appendages to serve the office of ambulatory locomotion, or of claws for the apprehension and tearing of the food, or of weapons of defence. In most of the Edriophthalma the normal arrangement obtains, and the thorax bears seven pairs of ambulatory members. The remaining appendages, which seldom exceed six pairs, belong to the abdominal portion of the body, and in the higher forms are very small and slightly developed, in comparison with those of the thoracic division. In the female Decapoda they constitute the support of the eggs, after their exclusion, and as long as they continue attached to the parent.

In their full development, each of these organs consists

of three distinct parts. The Stalk, which constitutes the essential part, and which is usually multiarticulate; the Palp, which is an appendage to the stalk, and ordinarily arises from its basal segment; and the Lash. It is not in all cases that these three portions exist, and in the Brachyura, for example, the foot-jaws are the only ones in which they are all present. The ambulatory thoracic legs in these are obviously composed only of the stalk, without either of the other members, and consist of six distinct joints. In the Macroura, however, the ambulatory feet, in some genera, have all the three elements; in others, one of them is wanting. Their modifications are almost innumerable, and often it would be impossible to distinguish their homologues, without extensive comparative examination.

It is impossible, in a mere sketch, introductory to a local Fauna, to enter, at any detail, into the various modifications now merely alluded to, but perhaps there is scarcely any group of animals in which the homologies are more recondite, the variations more interesting, and the relations between those variations and the habits and requirements of the animals more beautiful and instructive.

In order to give a general idea of the extent of these modifications, it may be stated that the ocular peduncles are the only appendages which are never devoted to any but their normal objects. The antennæ are, as has been before observed, sometimes modified into locomotive organs. The cephalic appendages about the mouth, the mandibles and maxillæ, are sometimes rudimentary, at other times they are modified into mere organs of apprehension. The thoracic members are sometimes locomotive organs, at others they subserve the nutritive function: the remaining thoracic members are, in some cases, prehensile, in others ambulatory, in others natatory, in others partially branchiophorous, and so on. The abdominal sometimes serve the purpose of swimming, at others of bearing and protecting the eggs, at others they are partially converted into branchiæ. Besides these modifications, some or other of them are, in many forms, either wholly wanting or rudimentary.

The digestive system appears under very various phases in the different groups of the Crustacea. The extremes of this diversity are found in those two primary divisions, the food of which is most opposite in its kind. In the one group, the whole of which are parasitic upon other animals, and which I have in this Introduction considered as belonging to a distinct class, the aliment consists of the juices of the creatures to which they are attached, and is obtained by suction. In these the normal elements of the organs for procuring or preparing the food for digestion are either rudimentary or wanting. In the higher forms of the true Crustacea, on the contrary, which subsist upon solid and often hard substances, and in many cases on living prey, the organs for pursuing, seizing, tearing, and comminuting the food, are carried to a high degree of development, and a corresponding difference is also found in the digestive organs themselves. The most elaborate condition of these organs is exhibited in the Decapoda, and especially in the Brachyura. It has been already stated that the appendages belonging to certain of the cephalo-thoracic segments are variously modified to serve their several offices; and in the latter order they have been shown to consist of six pairs, of which some are actual organs of mastication, as the mandibles or the true jaws, the foot-jaws or pedipalps generally serving to keep the food in contact with the former, whilst it is being broken up by them.

The buccal orifice in the Brachyura occupies the inferior face of the cephalic division of the body, and is bounded anteriorly by a crustaceous lamina of determinate form, which has been termed the upper lip, and posteriorly by another termed the lower lip. The mandibles occupy the sides of the opening. After these, and external to them, are the first, and then the second pair of true jaws, followed by the three pairs of pedipalps or foot-jaws, the last of which, when at rest, close the mouth, and include the whole of the preceding ones. In the Macroura, the pedipalps are very different in their forms, and have the aspect of very simple feet. In the Stomapoda they not only have the form, but the office also of the other locomotive organs, and hence the increased number of legs which appear to appertain to these, and especially to the Mysidæ. In the Edriophthalma, and the other lower forms, the parts about the mouth are fewer, and more simple. At the back of the mouth, a short œsophagus opens into the stomach, which is a very capacious cavity, occupying the whole depth of the body in the Decapods, and co-extensive with the gastric region of the carapace, already described. It is pretty distinctly divided into two portions, a cardiac and a pyloric, the former occupying the greater portion of the cavity, the latter of small dimensions.

The means of comminuting the food are not restricted to the complicated machinery above referred to, for the stomach itself contains a very remarkable apparatus, consisting of several hard calcareous pieces, which may be termed gastric teeth. These are attached to horny or calcareous levers fixed in the parietes of the stomach; they are moved by a complicated system of muscles, and are admirably adapted to complete the thorough breaking down of the aliment, which had already been to a considerable extent effected by the buccal appendages. These gastric teeth may be readily seen and examined in the larger species of the Decapoda, as in the large eatable crab and the lobster; and it will be readily perceived how perfectly the different pieces are made to act upon each other, and to grind the food interposed between them. Analogous structures, but of less complexity, are found in the Edriophthalma. The single and simple intestine extends in a direct line from the stomach, and terminates at the last segment of the abdomen. Immediately from its origin at the pyloric opening of the stomach, a notable enlargement is observed, but the rest of the canal is of uniform size. The enlarged portion is, in some cases, very short; in others, it occupies the larger portion of the total length.

The liver is of considerable volume in most of the families of Crustacea, and occupies in the Decapoda the greater portion of the visceral cavity. It consists of a mass of cæcal vesicles, ordinarily more or less elongated, and pouring the secretion into a system of membranous canals, the union of which forms ultimately a large trunk on each side, which opens into the pyloric portion of the stomach. Such is the structure of this important gland in the highest forms; but in the larger Stomapoda its structure is apparently granular, and it forms two series of lobes extending the whole length of the intestine,—and in the Edriophthalma, according to Prof. M. Edwards, it is reduced to "three pairs of biliary vessels, running alongside the intestine, the whole length of the body." There are other tubular appendages connected with the pyloric portion of the stomach, which are of considerable size in certain of the larger Decapoda, and which, from analogy, may with some probability be considered as pancreatic.

The respiration in this class is, with very few exceptions amongst the Isopoda, aquatic. In some of the lower forms, it would appear that there are no special organs devoted to this function, but in the higher these are very varied, and in many cases of a complicated character. The typical form of Crustacean respiratory organs may be considered that of lamellar branchiæ; and this form is found in the Decapoda, and particularly in the Brachyura; in the crab it is seen in its most complete development. The branchiæ are placed within a distinct cavity on each side, protected above by the carapace, and lying upon the upper surface of the thorax. They consist of a series of elongated pyramidal bodies, each composed of a vast number of plates or lamellæ, which are closely packed, but still admit of the free circulation of the water between them. The respiratory cavity has an afferent and an efferent opening, through which the water is propelled by a mechanism differing in the different groups. The former opening, through which the water has access to the cavity, is a long lateral slit, between the cephalo-thorax and the side of the thorax; and the latter is near the buccal cavity, and is covered by the last or flabelliform appendage of the second pair of the true jaws, which is developed into a broad horny plate, fixed by a sort of pivot, on which it continually turns, and thus regulates the efflux of the water. Prof. Milne Edwards observes, that this action is proved to be essential to the renewal of the water which bathes the branchiæ, as, if its movements be stopped, the animal becomes soon asphyxiated. The whole of the apparatus belonging to this function in the higher Crustacea is exceedingly curious and interesting, but it would be out of place to enter into the detail in this work.

The branchiæ are very differently formed in the different orders of the class, and even vary considerably in some genera of the same family. In some cases the abdominal appendages support these organs; in others they are attached to the basal joint of the thoracic legs; in some genera, as in Mysis, their distinct existence has not as yet been demonstrated, although, as I have observed in speaking of that genus in the body of the work, there appears little doubt that a special organ exercises their function.

In the terrestrial Isopoda, or the common Millipedes, as they are termed, the respiration is exclusively atmospheric.

The respiration of the land crabs, which must necessarily be, during the greater part of their lives, atmospheric, is one of the most remarkable phenomena connected with this subject, and has occupied the attention of Mons. Audouin and Dr. Milne Edwards, who have given a most elaborate and interesting memoir on this subject,* to which the reader is referred. It is well

* Annales des Sciences Naturelles, t. v. p. 85.

known that the lobster will live for a long time out of water, provided the branchiæ are occasionally bathed, so as to keep them in a humid condition, whilst it will die very soon on being confined in a small quantity of water, without access to air.

There has been considerable discrepancy in the statements of different anatomists respecting the circulation in the Crustacea. Messrs. Audouin and Milne Edwards* have considered that "no other than the two great branchial veins terminate in the heart, and, consequently, only pure aërated or arterial blood is propelled by it over the general system; the circulation is, in fact, the same as in the Gasteropodous Mollusca; the ventricle is exclusively systemic, and is provided with only two venous apertures." Such is a summary of their opinion. The fact, however, that the circulation is of a mixed kind was evidently known to Hunter, and has been elaborately demonstrated by Professor Owen in his more recent researches.† A reference to the engravings from the Hunterian drawings in the collection of the Royal College of Surgeons,‡ to that of the heart of the lobster by Professor Owen in his lectures above referred to, and to the respective descriptions of these figures, will show "that the heart, instead of being purely systemic, is partly branchial, and impels the blood, not through the body only, but also to the respiratory organs."

* Recherches Anatomiques et Physiologiques sur la Circulation dans les Crustacés. Ann. des Sc. Nat. t. ii.
† Lectures on the Comparative Anatomy and Physiology of the Invertebr.
‡ Catalogue of the Physiological Series of Comparative Anatomy contained in the Museum of the Royal College of Surgeons, vol. ii. Copied in Professor Rymer Jones's "Animal Kingdom," pp. 333-336.

"We may trace," says Professor Owen,* "in the heart of the Crustacea, a gradational series of forms, from the elongated, median, dorsal vessel, to the short, broad and compact muscular ventricle in the lobster and the crab. In all the Crustacea, as in all the other articulate animals, the heart is situated immediately beneath the skin of the back, above the intestinal tube, and is retained in situ by lateral pyramidal muscles. In the lower, elongated, many-jointed species of the Edriophthalmous Crustacea the heart presents its vasiform character: it is broadest and most compact in the crab. In this series we may trace a general correspondence in the progressive development of the vascular as of the nervous system, concomitant with the concentration of the external segments, and the progressive compactness in the form of the entire body."

Corresponding with the view which has been taken of the gradual condensation of the segments of the body and the centralization of the viscera, is that of the nervous system as seen in the various forms of Crustacea as they rise in the scale of organization. An elaborate detailed description of all the gradations formed the substance of an admirable essay † by the distinguished naturalists so often quoted, of whose labours an excellent abstract is given by my friend Professor Rymer Jones, in his "Animal Kingdom." ‡

In Talitrus, where the insectiform arrangement is the most obvious, and where every pair of ganglia consists of

---

* L. c. p. 180.

† Messrs. Audouin et Milne Edwards, "Recherches Anatomiques sur le Système Nerveux des Crustacés." Ann. des Sc. Nat. t. xiv.

‡ L. c. p. 337.

two separate nuclei of nervous substance, united by a transverse band, with an anterior and posterior nervous filament uniting each to the antecedent and succeeding pairs, the number of ganglia (thirteen) coincides with that of the segments of the body. Proceeding upwards, a condensation, both lateral and longitudinal, of certain of the ganglia is found to be coincident with the concentration of the rings, until in the crab the whole of the abdominal and thoracic ganglia become concentrated into one mass, from which the nerves radiate in a most beautiful manner to the parts about the mouth, the limbs, &c. The conclusions to which their elaborate researches have conducted Messrs. Audouin and Milne Edwards are thus given :—

"Le système nerveux des Crustacés se compose toujours de noyaux médullaires dont le nombre normal est égal à celui des membres, et toutes les modifications qu'on y rencontre dépendent principalement de rapprochemens plus ou moins complets de ces noyaux, agglomeration qui s'opère des côtés vers la ligne médiane en même temps que dans la direction longitudinale, mais peuvent tenir aussi en partie à un arrêt de développement dans un certain nombre de ces noyaux." *

The organ of hearing is found only in the higher forms of this class. In the larger Decapoda, and particularly in the brachyurous group, it is very easily seen, on removing a little crustaceous plate in the basal joint of the second antennæ, and thus exposing a small cavity. This operculum is pierced by a small oval opening, covered with a membrane; and in the Macroura, the whole closure

---

* Hist. Nat. des Crustacés, t. i. p. 147.

is membranous. Within the cavity and immediately behind the little opening before mentioned, is a minute vesicle filled with fluid, which conveys the vibrations to a branch of the antennal nerve, which is expanded upon the vesicle. This is the simple apparatus; but it is sufficient to receive and convey to the sensorium the imperfect sonorous vibrations to which they are subject.

The visual organ is essentially similar to that of insects. The eyes are compound in all the higher forms, and those of the Edriophthalma do not differ essentially from those of the Podophthalma, excepting in the absence of those movable peduncles by which the eyes of the latter are distinguished. The optic nerve, the lenses, the facets of the cornea, the pigment, are alike in all, and in all resemble generally the same organs in insects. There is one peculiarity, however, which is found in certain species which live in such places as are inaccessible to light, or to such degrees of it as would render eyes in any way useful. In *Calocaris*, for instance, a little prawn-like animal, inhabiting very deep water, and ordinarily immersed in mud, the eyes and their peduncles do not differ in form from those of the other Palæmopidæ; but the vision is wanting. There is no pigment, there are no corneal facets; the organ is evidently rudimentary and merely formal. Mr. Westwood has recently made known through the Linnean Society a form of Edriophthalma, inhabiting a deep well, a species in which there is no external appearance of eyes whatever; Mr. Newport has, however, by his accustomed accuracy of dissection, shewn that in this case also, a rudimentary visual organ exists underneath the cephalic crust.

The propagation of the Crustacea proper is invariably oviparous, and the sexes are distinct. The reproductive organs in either sex are double, the two elements being perfectly similar, and occupying a corresponding position on each side of the median line. The two are wholly independent of each other, having no communication even to the efferent opening, there being one of them to each. Dr. Milne Edwards mentions the following curious fact:—" Cette indépendance des deux moitiés de l'appareil de génération est si complète qu'on a vu un cas, où l'un des côtés était mâle et l'autre femelle, sans que cette monstruosité eût entrâiné aucune autre perturbation sensible dans la conformation de ces organes."* They are very similar in arrangement, position, and general relation to the other organs in the two sexes.

In most cases the eggs are carried by the female until they are hatched; but in some they are previously deposited in the sand. In different families the eggs are carried by the mother attached to different parts of the body. In the Decapoda they are borne on the under side of the abdomen, attached to the abdominal false feet. In the genus *Mysis*, a pouch is formed at the base of the posterior thoracic legs,† in which the eggs remain until the young are excluded. In *Thysanopoda*, another genus of the Mysidæ, they are contained in two oval purses, depending from the same part.‡

* Hist. des Crust. t. i. p. 165.   † See p. 336.   ‡ See p. 346.

ON EXUVIATION AND THE RESTORATION OF LOST LIMBS.

The fact that the throwing off of the old integument and its replacement by a new one during the growth of the animal, takes place in all the Crustacea as necessarily and as constantly as in insects during their larva condition, has long been known, and as long has excited the admiration of all who take any interest in natural phenomena. That an animal covered by integument of the hard, solid, almost stony consistence as that of the lobster and the crab, for example, should have the power of withdrawing itself from its shell, leaving it, to all appearance, as perfect as before, with the carapace, the abdomen, the limbs, the eyes, the antennæ, and even the stomachal teeth, and other internal shelly organs, whole and entire, and in their former relative situation and condition, is one of the most interesting, and, at first sight, one of the most perplexing and inexplicable, of all the phenomena of voluntary action.

The first clear and satisfactory observations on this subject were made by Rëaumur,[*] whose unexampled accuracy and truthfulness is attested by the fact that of all the observations made by himself alone, far exceeding those of any other naturalist of past or present times, and occupying, in their published form, numerous large quarto volumes, scarcely one has ever been contravened by subsequent credible observers, whilst they have formed the substance of half the numerous compilations on insect life, acknowledged or otherwise, which have appeared since his time.

[*] Mém. de l'Acad. des Sc. 1712, p. 226, and 1718, p. 263.

The necessity for the process in question is so evident, seeing that, without it, there would be no possible means of allowing the gradual growth of the animal, that it is matter of surprise that it should have ever been doubted, as it appears to have been by a distinguished entomologist, more especially of late years, when so many conclusive observations have been made of the fact. There is no doubt that in many of the higher forms it takes place annually, with great regularity,* until the growth is completed, which in many species is not the case before the animal is many years old. This is proved by the extent to which the size increases by each moult, compared with the difference between the young and the old animal; and it is evident that after the growth has reached its maximum the crust ceases to be changed, from the fact which I have seen in several instances, as in the common crab, the lobster, and some others, where the carapace of the still living creature was the seat of barnacles so large, that several years must probably have been required for attaining their existing size.

The observations of Rëaumur to which I have alluded, and those of subsequent naturalists, and especially of Mr. Couch, furnish us with the following history of this curious process.

When the animal by gradual internal increase has become too large for its existing covering, it ceases for a time to feed, and retires to a secret and undisturbed situation, where it may undergo the process in security. If it be examined at this time, an evident loosening of the

---

* Some recent observations by Mr. Warrington shew that in the common prawn, the moult is much more frequent; he has noticed its occurrence with much regularity, every twelve days, in the summer.

crust may be perceived, upon pressing it gently in different parts. Shortly afterwards,—and this description belongs particularly to the river cray-fish,—it appears uneasy and restless, rubbing its limbs against each other, and moving the segments of the body in various directions. It throws itself on its back, and, swelling out its body, ruptures the membrane which connects the carapace with the abdomen, and raises the former, so as to loosen it from its attachments. Resting from time to time after its laborious efforts, it finally detaches the whole thoraco-abdominal portion, from which it withdraws itself, having, with much apparent difficulty and pain, disengaged the legs, and then the antennæ, the eyes, and other appendages. It is impossible to imagine that the crust of the legs, and especially of the great claws of the larger species, could be cast off unless it were susceptible of being longitudinally split; and Rëaumur states that such is actually the case; each of the segments being composed of two longitudinal pieces, which, after separating to allow of the passage of the soft limb, close again so accurately that it is very difficult, in the cast crust, to discover the line of division. When the animal has disembarrassed itself of the crust, the latter is found absolutely entire, and has exactly the form which it possessed previous to the operation. In a recent interesting account of the exuviation of a Maia,* Mr. Gosse has, however, shewn that in this brachyurous form, no such splitting of the legs takes place, but that "the animal pulled first at one and then at another, until they were quite out, as if from boots. The joints as they came out were a great deal larger than the cases from which they proceeded. It

* Annals of Nat. Hist. 2nd Ser. vol. x. p. 210.

was evident that, in this instance, neither were the shells split to afford a lateral passage for the limbs, nor were the limbs reduced to tenuity by emaciation." The new integument is at first soft and membranous, but speedily becomes encrusted with calcareous matter, and as hard as the former. The additional size which is gained by each moult is very striking, and I have often felt, on seeing a newly-emancipated crab by the side of the shell which it had just shed, that, were not the fact absolutely ascertained by observation, it would appear physically impossible that the larger body could have so recently been contained within so small a case. Rëaumur supposed that even the hairs with which the surface is in many species furnished, were contained within the cast crust; but Dr. Milne Edwards asserts that such is not the case; stating that they are not at first obvious on the surface of the new shell, but " sont rentrés à l'intérieur, comme le doigt d'un gant qui serait retourné sur lui-même!" If we open, says this author, a Maia a short time before the commencement of the moult, we find between the existing shell and the " chorion" a membranous layer, which resembles condensed cellular tissue, and which becomes thicker and more solid, as the period of moult approaches; it is evidently secreted by the chorion, and is moulded upon the shell which covers it. In the common crab (*Cancer Pagurus*), and some others of similar form, it would appear that the carapace, instead of being cast entire, divides at the junction of the epimera with the dorsal piece or tergum; a fact which I have often seen in many species, particularly in the larger Grapsidæ, which, from their form, could not possibly withdraw the body without such a separation.

In the account of the great crab, p. 62, I have stated that the male lies in wait for the female previous to and during her moult, and seizes her as soon as this is accomplished, whilst she is still weak and enfeebled by the process; and I have so commonly seen the male and female shore-crab (*Carcinus Mænas*) in conjunction when the latter is still soft, that there can be no doubt that this is a general, although certainly not a constant habit.

A no less curious and interesting process than that above described, is the voluntary casting of the limbs, and the restoration of such as have been thus lost by the animal's will, or by accident. Rëaumur in this case also was the first to make any correct and scientific researches on the subject, and his statements, full of interest, will be found in the earlier of the two memoirs already quoted. My friend Mr. Couch has subsequently extended these observations, which will be found embodied in my account of the habits of the lobster at page 245.

On this subject an interesting paper was read before the Wernerian Society of Edinburgh by Mr. H. Goodsir, in December, 1843; and to the details which I have given in the place above mentioned, I would merely add a short abstract of Mr. H. Goodsir's paper:*

" It has long been known that the animals belonging to this class have the power of reproducing parts of their body which have been accidentally lost. If one of the more distant phalanges of a limb be torn off, the animal has the power of throwing the remaining part of the limb off altogether. This separation is found to take place always at one spot only, near the basal extremity of the first

* Annals of Nat. Hist. vol. xiii. p. 67.

phalanx. The author has found that a small glandular-like body exists at this spot in each of the limbs, which supplies the germs for future legs. This body completely fills up the cavity of the shell for the extent of about half an inch in length. The microscopic structure of this glandular-like body is very peculiar, consisting of a great number of large nucleated cells, which are interspersed throughout a fibro-gelatinous mass. A single branch of each of the great vessels, accompanied by a branch of nerve, runs through a small foramen near the centre of this body, but there is no vestige of either muscle or tendon, the attachments of which are at each extremity. In fact, this body is perfectly defined, and can be turned out of the shell without being much injured.

"When the limb is thrown off, the blood-vessels and nerve retract, thus leaving a small cavity in the new-made surface. It is from this cavity that the germ of the future leg springs, and is at first seen as a nucleated cell. A cicatrix forms over the raw surface caused by the separation, which afterwards forms a sheath for the young leg."

METAMORPHOSIS.

One of the most marked characters by which this class was long considered as distinguished from that of insects, was the supposed absence of any such change of form, during the progress of development after exclusion from the egg, as is ordinarily understood by the term *metamorphosis;* and Dr. Leach, in his definition of the class,* formally adopts this character, which has been repeatedly recognised by others.

* Encycl. Brit., Art. Crustacea.

It was in the year 1823 that Mr. Vaughan Thompson, whose name is now identified with the discovery, following up an observation made by Slabber, a Dutch naturalist, as long ago as 1768, and published ten years afterwards, established the remarkable fact that those anomalous forms which constituted the genus *Zoea* of Bosc, are nothing more than the early or larva condition of the higher Crustacea. It will readily be imagined that no small excitement was produced in the scientific world by the announcement of a discovery which, followed up, as it afterwards was, with equal intelligence and perseverance, and with corresponding success, may claim for its author a place amongst the few observers who, from a single phenomenon, have been led to the establishment of generalisations and laws of the highest importance.

Notwithstanding, however, the credit is due to Mr. Thompson of having carried out the suggestion to its full development, it was undoubtedly to the Dutch naturalist that he was indebted for the ascertained fact that the anomalous creatures on which Bosc afterwards founded his genus *Zoea* pass by metamorphosis into a different and a higher form.

Before I proceed with the further history of this discovery, I think it right to show the grounds of Slabber's claim, which had been wholly overlooked as to its results, and which, in consequence of an error arising from deficient information, Mr. Thompson himself, in the first place, much depreciated, without, as far as I am aware, having afterwards taken any opportunity of correcting the misapprehension. It was, then, in the year 1778 that Slabber published a small work, in which occurs a description with figures of a new crustacean animal

Fig. a.

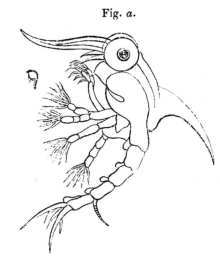

Fig. b.

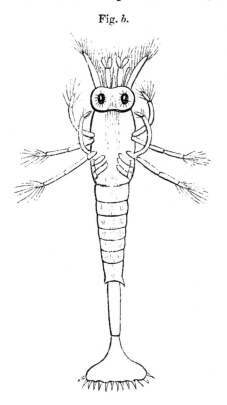

(fig. *a,*) to which the name of *Zoea Taurus* was afterwards given. Having taken at sea several specimens of this singular creature, he placed one of them (*a*) in sea water, which he constantly renewed, for the purpose of observation, and, " on the third day, finding its movement become slower and its colour paler, he subjected it to the microscope, and found to his surprise that the anterior part of the animal had changed its form, and on the fourth day it had acquired the appearance represented in fig. *b*, so that, together with the other individuals he had taken, it seemed to have experienced a complete metamorphosis; under this new form the dorsal spine had disappeared, the front spine had become comparatively small, the antennæ were rendered conspicuous, the feet and eyes were apparently more developed, and the tail had changed from forked to spatulate, fringed by a row of thirteen short spines." It would certainly seem that this plain and simple statement, supported as it was in many respects by Mr. Thompson's own subsequent observations, can scarcely justify the conclusion to which that gentleman is led,* "that Slabber lost his Zoea, in changing the sea water, and that the new form came from the added portion." But the truth of Slabber's statement, and, consequently, the evidence of the correctness and originality of his discovery, are very strongly proved by the almost absolute identity of the second form of his animal with that of several species subsequently observed; and particularly of the ditch-prawn, *Palæmon varians*, as figured by Capt. Du Cane.†

It was, however, from this observation of Slabber that Mr. Thompson, in the year 1823, was induced to carry out the investigation. In the spring of the previous year, as he informs us, in the harbour of Cove, he first met with Zoeas, and that in considerable abundance; and " in the year following, at the same season, one of considerable size occurred, amongst a number of smaller ones, and, judging it full grown, he considered it a fit subject to keep for the purpose of witnessing the *metamorphosis observed by Slabber*," &c. The metamorphosis was interrupted by the death of the animal when in the act of undergoing it; but it had advanced sufficiently to show that the animal belonged to the Brachyura, and the portion which was observed, contained all the five feet on one side, the anterior foot being furnished with a perfect claw; and it appears now more than probable that the form into

* Zool. Researches, p. 8.   † Ann. Nat. Hist. vol. ii. pl. 6.

which it was passing was that of *Megalopa,* to which further reference will be presently made.

Here, then, was the first decided demonstration; but any doubt which might be supposed to appertain to an incomplete fact, was shortly removed by Mr. Thompson's success in hatching the ova of the common crab, *Cancer Pagurus,* the product of which were true Zoeas.

Subsequent observations by Mr. Thompson confirmed his new views, and he established the truth of a metamorphotic change in several genera; the results of his researches being given to the world in a subsequent portion of his Zoological Researches, in the " Entomological Magazine,"* in " Jameson's Journal,"† and particularly in a paper read before the Royal Society in 1835, and published in the " Philosophical Transactions," in which details are given of the complete changes in *Carcinus Mœnas,* the common shore crab, which establish the further interesting and important fact, that while the animal appears under the aspect of a *Zoea* on its first exclusion from the egg, it undergoes a further change into a true *Megalopa* before its final assumption of the perfect form: showing that this supposed genus also, which was formed by Leach, is, like Zoea, only a phase of a higher type. Thus, in its progress from the egg to its final development, the brachyurous crustacean was proved to pass through two temporary conditions, which had previously been regarded as types, not of genera only, but of different families; and both strikingly dissimilar from the group to which, in its perfect state, it really belongs.

The new doctrine was not received at once with implicit assent. Mr. Westwood, in a paper read before the

* Vol. iii. pp. 85, 275, 370, 452. † For 1846.

Royal Society in June, 1835,* not only contests the universality of the law, which Mr. Thompson had somewhat too hastily, perhaps, deduced from his facts, but concludes that that gentleman's views are erroneous, and that "*no exception* occurs to the general law of development in the Crustacea—namely, that they undergo no change of form sufficiently marked to warrant the application to them of the term metamorphosis."

This hasty, and, as the result has proved, very premature condemnation, derived some *primâ facie* supports from the elaborate investigations of Rathke on the development of the embryo in the ova of the river cray-fish, *Astacus fluviatilis,* and the subsequent observations of Mr. Brightwell on that of the lobster, which latter, however, have since been only partially verified by Rathke, and are, indeed, modified in some particulars by Mr. R. Couch. To these I shall have occasion to refer more particularly hereafter ; it is sufficient now to observe, that in both instances the animal was stated to be perfected by gradual development, and not by any sudden change of form. These, if even the statements were fully borne out, have since been proved to be merely exceptional cases; and not only is Mr. Rathke's assumed general support of Mr. Westwood's objections completely removed, but that distinguished physiologist himself volunteers his strong testimony in favour of the opposite views in a subsequent paper, in which he says that he hastens the publication of these new researches respecting the development of several other forms of Crustacea, *one of which is the lobster,* " in order, as soon as possible, to record a testimony to the correctness of Thompson's dis-

* Phil. Trans. 1835, p. 311.

covery, that even the Decapods, after they have already quitted the egg, undergo a very considerable metamorphosis;" and, in conclusion, he adds, " from the notices which I have here briefly communicated respecting the development of some Decapods, it results that several of these animals, as first discovered and described by Thompson, undergo a very considerable and highly remarkable metamorphosis. . . . . I, therefore, confess that I have done Thompson injustice in not putting faith in that discovery." And he then states his intention " next spring, partially to subject his researches on the cray-fish to revision."* There is one apparent anomaly, however, on which Mr. Westwood dwells with some plausible show of reason, and on which it may be well to offer a few remarks.

Amongst the specimens of Crustacea, preserved in spirits, which formed part of the collection of the late Rev. Lansdown Guilding, and which came into my possession after his death, was one of the abdomen of a female land crab, *Gecarcinus,* to which were attached numerous young, in their perfect form, and very similar, excepting in size, to the parent. Here, then, was a case in which, it may at once be granted, no *external* and independent metamorphosis, at least, had taken place; and on this, with the other instances above alluded to, Mr. Westwood founds his principal argument against the doctrine enunciated by Mr. Thompson. But may not this probably be an analogous phenomenon to that of the land salamanders amongst the amphibia? And, as in that instance, where the parent has no opportunity

* Wiegmann's Archiv. part iii. 1840. Translated in Ann. Nat. Hist. vol. vi. pp. 263-268.

of depositing her eggs in the water, where, in the more typical forms, the young undergo the transformations essential to the whole group, the changes take place in the oviduct; so may not the young of the land crab, whose habits require them to be speedily in a condition to leave the coast where they are hatched, formally undergo the metamorphosis within the egg? This being granted, it would be as reasonable to deny the phenomenon of transformation in the amphibia generally, because the young of the salamander are brought forth in the perfect state, as to deny its occurrence in the Crustacea, on the analogous exceptional case of the terrestrial *Gecarcinus*.*

I do not consider it necessary to examine at any detail the "six arguments" which Mr. Westwood adduces "against the metamorphosis into crabs which the Zoes are stated to undergo," since the facts, exactly as related by Mr. Thompson, have been so fully confirmed by subsequent observers. Indeed, I prefer referring to the whole of Mr. Westwood's elaborate examination of the question, for the information of those who may have the curiosity to see how much may plausibly be urged against the truth of a theory, so irrefragably supported by facts. It is sufficient to say that Mr. Westwood does not attempt to bring forward a single investigation or observation of his own in support of his views, with the exception of that of the land crab, already mentioned.†

---

* Mr. Thompson, in the case of *Gecarcinus*, as in that of some other West-Indian species, depended for his information upon some specimens of female crabs with matured ova being sent to him in spirits. The ambiguous character of such observations may warrant us in eliminating them at once from the question.

† I have thought it necessary to examine Mr. Westwood's objections at

But Mr. Westwood was not the only one who demurred to the correctness of Mr. Thompson's conclusion. In the first volume of Milne Edwards's admirable "History of Crustacea,"* this author says, " Les Decapodes paraissent tous naître avec la série complète de leurs anneaux et leurs membres ;" and in a note occurs the following opinion on the earliest researches of Mr. Thompson. " Suivant M. Thompson, les Decapodes éprouveraient de véritable métamorphoses, car ce naturaliste regarde l'animal connu sous le nom de Zoé comme étant le jeune du crabe commun de nos côtes. Mais cette opinion n'est pas étayée d'observations assez précises pour entraîner la conviction."

It is remarkable that this distinguished naturalist's ultimate convictions were derived from his own observation; and it is difficult to account for such a discrepancy when we consider the high character of the dissentient, and the means which were placed in his hands for determining the question; for in consequence of the interest which it excited amongst the scientific men of France, Dr. Milne Edwards was deputed with another naturalist, to repair to the Isle de Rhé for the express purpose of settling the disputed point, and he arrived, as we learn, at the conclusion above stated.

some length, on account of that gentleman's deserved eminence as a profound entomologist, and because I believe that he has never published any recantation of the opinions stated in his paper. I have, however, before me, a letter from him to myself, dated Sept., 1844, in which the following passage occurs, showing that his convictions on this subject had undergone a material change :—" I believe it will turn out, following the normal rule of development of the embryo, that at a certain period all the Decapods are Zoeæ, and that some are born (*i. e.* escape from the egg) in that state, but that others are not born until a late period of development, that is, when the true legs and claws are disengaged."

* P. 198.

Subsequently to the researches above-mentioned, the late Capt. Du Cane investigated the development of the shore crab, *Carcinus Mænas,* and of the Ditch Prawn, *Palemon varians,* with complete success; establishing in each of these forms the truth of Mr. Thompson's position. Mr. H. Goodsir also examined, with similar results, the former species. But by far the most complete illustration of the subject and the most extensive proofs of the general law, are afforded by the researches of my friend, Mr. Richard Q. Couch, of Penzance, who, dissatisfied with the uncertainty and contradiction of former testimony, resolved to investigate the matter for himself; and this he effected with a degree of acumen and perseverance which characterise all his researches, and by which the truth of the doctrine is fully established, as regards the genera *Cancer, Zantho, Pilumnus, Carcinus, Portunus, Polybius, Maia, Galathea, Homarus,* and *Palinurus*—a goodly number to have been investigated by one observer—and of some of these he watched every change. These results were published in two Memoirs, read to the Cornwall Polytechnic Society in 1843; in which the author takes a clear and fair view of the whole subject, and comes to his decision with a host of evidence sufficient to set the substantive question entirely at rest. Unfortunately, the useful local publication in which these memoirs appeared, is so much confined in its circulation that it has probably fallen into the hands of but few naturalists.

I have felt it desirable to give a more extended history of the discovery, as, with the exception of Mr. R. Couch's first memoir just referred to, no such digest has ever been placed at one view before the world. I now proceed to examine the actual results, and to endeavour to reduce the facts already known to some order.

It will be inferred from the previous account, that there are considerable variations in the character of the metamorphosis of different families, and that in the case of *Astacus fluviatilis,* there appears at present to exist even an abrupt and isolated exception to the general law. As this is the only case at present in which such exception has been established, I refer my readers for further information on this subject to the work of Mr. Rathke himself,* which constitutes one of the most complete and elaborate monographs in existence, illustrated in the most beautiful and perfect manner; and to the full and satisfactory analysis of the work by Milne Edwards in the first volume of his " History of the Crustacea."

Eliminating, therefore, this exceptional case, it will be found that the fact of a metamorphosis has been demonstrated with more or less success in no less than seventeen genera of the Brachyurous order of the Decapoda —in which order the phenomenon is most decided and obvious—belonging to the families *Leptopodiadæ, Maiadæ, Canceridæ, Portunidæ, Pinnotheridæ, Grapsidæ,* and *Gecarcinidæ.* In the Anomourous order, it has been shown in the genera *Pagurus, Porcellana,* and *Galathea,* and amongst the Macroura in *Homarus, Palinurus, Palemon,* and *Crangon.*

The facilities which everywhere exist for procuring the common shore crab, *Carcinus Mænas,* have occa-

---

* Untersuchengen neber die Bildung und Entwickeberg des Flusskrebses, von Heinrich Rathke. Folio. Leipzig. 1829.

sioned it to be more fully investigated than any other; and it may, therefore, be taken as the type of the process amongst the Brachyura. Thus it was the first form in which the Megalopoid period was observed by Mr. Thompson;* it was four years afterwards described in its zoeform state by Capt. du Cane, who, it appears, was not acquainted with Mr. Thompson's paper; it has occupied the attention of Mr. H. Goodsir; and it forms the subject of Mr. R. Couch's elaborate and very complete researches. To the latter of these, as embodying all that is at present known on the subject, and as being the result of the personal observation of so intelligent and acute an observer, I shall have recourse for the general description of this process in the Brachyura. In the first place it appears that Mr. Couch met with the young Zoes already hatched; and even then he had the satisfaction of finding them pass into the Megalopoid condition described by Thompson. Afterwards, however, he procured some specimens of the crab itself laden with ripe ova, just ready for shedding; and he then proceeds with the account of his observations:—

"These were transferred to captivity, placed in separate basins, and supplied with sea water, and in about sixteen hours I had the gratification of finding large numbers of the creatures alluded to above, swimming about with all the activity of young life. There could be but little doubt that these creatures were the young of the captive crabs. In order, however, to secure accuracy of result, one of the crabs was removed to another vessel, and supplied with filtered water, that all insects might be removed; but in about an hour the same crea-

* Phil. Trans. ut supra.

INTRODUCTION.

tures were observed swimming about as before. To render the matter, if possible, still more certain, some of the ova were opened, and the embryos extracted; but shortly afterwards I had the pleasure of witnessing, beneath the microscope, the natural bursting and escape of one precisely similar in form to those found so abundantly in the water. Thus, then, there is no doubt that these grotesque-looking creatures are the young of the *Carcinus Mænas;* but how different they are from the adult need hardly be pointed out any further than by referring to the fig. (*c*). When they first escape they rarely exceed half a line in length. The body is ovoid, the dorsal shield large and inflated, on its upper edge and about the middle is a long spine, curved posteriorly and rather longer than the diameter of the body, though it varies in length in different specimens; it is hollow, and the blood may be seen circulating through it. The upper portion of the body is sap-green, and the lower semi-transparent. The eyes are large, sessile, and situated in front, and the circumference of the pupil marked with radiating lines. The lower margin of the shield is waved, and at its posterior and lateral margin, is a pair of natatory feet. The tail is extended, longer than the diameter of the shield, and is composed of five equal annulations, beside the

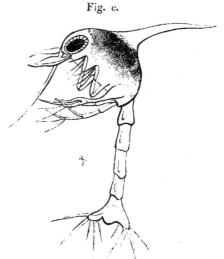

Fig. *c*.

terminal one; its extremity is forked, and the external angles long, slender, pointed, and attached to the last annulation by joints. Between the external angles, and on each side of the median line, are three lesser spines, also attached to the last ring by joints. Between the eyes, and from near the edge of the shield, hangs a long, stout, and somewhat compressed appendage, which, as the animal moves, is reflexed posteriorly between the claws. Under each eye there is also another appendage, shorter, and slightly more compressed. The claws are in three pairs; each is composed of three joints, and terminates in four long, slender, hair-like appendages. These claws are generally bent on the body, but stand in relief from it. If the animal be viewed in front, the lower margin of the dorsal shield will be found to be waved into three semicircular festoons, the two external of which are occupied by the eyes, and between which the middle one intervenes; the general direction of the claws will be seen to be at right angles to the body. As the young lies enclosed within the membranes of the egg, the claws are folded on each other, and the tail is flexed on them so far as the margin of the shield, and, if long enough, is reflected over the front of the shield between the eyes. The dorsal spine is bent backwards, and lies in contact with the dorsal shield; for the young, when it escapes from the egg, is quite soft, but it rapidly hardens and solidifies by the deposition of calcareous matter, in what may be called its skin. The progress of this solidification may be very beautifully observed by watching the circulation in the dorsal spine. When the creature has just effected its liberation from the egg, the blood globules may be seen ascending to the apex; but as the consolidation advances, the circulation becomes more and more limited in its extent, and is finally confined to the base. These minute creatures, in this early state of their existence, are natatory, and wonderfully active. They are continually swimming from one part of the vessel to the other, and when observed free in their native pools, if possible even more active than when in confinement. Their swimming is produced by continued flexions and extensions of the tail, and by repeated beating motions of their claws; this, together with their grotesque-looking forms, gives them a most extraordinary appearance when under examination. As the shell becomes

Fig. d.

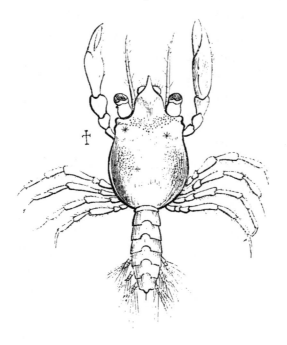

more solid they get less active, and retire to the sand

at the bottom of the vessel, to cast their shells, and acquire a new form. They are exceedingly delicate, and require great care and attention to convey them through the first stage; for unless the water be supplied very frequently and in great abundance, they soon die. The second form of transmutation is equally as remarkable as the first, and quite as distinct from the adult animal (*d*). In the species now under consideration this second transformation is marked by the disappearance of the dorsal spine; the shield becomes flatter and more depressed, the anterior portion more horizontal and pointed, the three festoons having disappeared. The eyes, from being sessile, are now elevated on footstalks; the infra-orbital appendages become apparently converted into antennæ. The claws undergo an entire revolution; the first pair become stouter than the others, and are armed with a pair of nippers," the others being simple; " but the posterior pair are branched near the base, and one of the branches ends in a bushy tuft. The tail is greatly diminished in its relative size and proportions, and is sometimes partially bent under the body, but is more commonly extended. This form is as natatory as the first. They are frequently found congregating around floating sea-weed, the buoys and strings of the crab pot marks, and other floating substances, both near the shore and in deep water. Their general form somewhat resembles a Galathea."

Every one will immediately recognise in this description, and in the figure which accompanies it, the creature typifying the genus *Megalopa* of Dr. Leach. Here, then, is the second form of a brachyurous type, and its final change is seen in the accompanying figure (*e*). It is

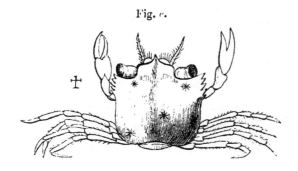

Fig. *e*.

unnecessary to follow out the minor distinctions in the various brachyurous genera. It is sufficient to state that the investigations of Mr. Couch confirm fully the views of Thompson, by the establishment of a metamorphosis of similar character, as regards the first change, in the large edible crab, *Cancer Pagurus*, in *Portunus*, and in several of the *Maiadæ* and *Leptopodiadæ*. There are some minor differences in the structure and form of the first stage of these as compared with that of *Carcinus*, but they do not involve any important consideration. The curious little larva of *Pinnotheres* I have figured at p. 125, after Thompson, and as I have myself seen it.

Amongst the oxyrhynchous forms there are some rather curious deviations from the type above described, particularly in the absence, according to Mr. Couch's figures in the genus Maia, of the dorsal and frontal spines; but these, as I understand Rathke's description, are found in the corresponding stage of the neighbouring genus *Hyas;* if this be so, it shows that the existence or absence of these spines is of little importance.

In the Anomoura we have elaborate descriptions of the young stage of *Pagurus*, in the paper by Rathke already referred to, and in one by Dr. Philippi, with a figure.*

* Ann. of Nat. Hist. vol. vi. p. 92, pl. iii. f. 7, 8.

If this figure be correct, we have a remarkable approach in the general form of this species to that of some of the smaller Macroura, as observed by Mr. Thompson and Capt. Du Cane; but the details scarcely agree with the full and doubtless correct description of the former author. The researches of Mr. Rathke * are, in fact, of great value, as affording the only clue we have yet seen, to the homologies of the members which exist in this early condition of the animal. It appears from this account, that the true feet are not represented by the three pairs of locomotive organs which are observed in the early stage, but that these are in fact developed into the foot-jaws of the adult. " Embryos about to escape have only three pairs of members that can serve for locomotion. All these six members are not, as might be expected, true feet in a lower state of development, but the foot-jaws. Of true legs, and also of branchiæ, there does not yet exist a trace." It is not until a subsequent period that these organs are formed, and, in fact, the whole account of the development of the young *Paguri*, as given by M. Rathke, is highly interesting, and would be particularly useful as a guide to those observers who might have the opportunity of watching the whole progress of any of these animals from the egg to maturity.

The most remarkable form of the larva amongst the Anomoura hitherto observed, and, indeed, one of the most anomalous in the whole Decapod group, is that of *Porcellana platycheles*, as described and figured by Mr. R. Couch, in his second Memoir. There is no appearance of either dorsal or frontal spines, in which respect it agrees with the Macroura, as it does also in the com-

* Ann. of Nat. Hist. vol. vi. p. 263.

lvi INTRODUCTION.

pressed corslet, and the large, sessile eyes. On its first escape from the egg (fig. *f*), the feet are in two pairs, dichotomously branched and destitute of hairs; the tail comparatively short, the terminal flap somewhat lozenge-shaped, and armed with long, slender, bristle-like appendages. From the anterior part of the carapace hang two long, slender filaments which turn under the thorax. In a few hours the first exuviation takes place, and the animal appears under a different aspect (fig. *g*). The

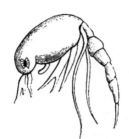

Fig. *f*.

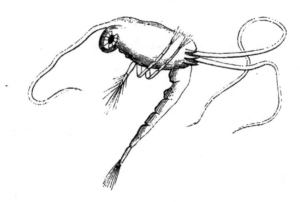

Fig. *g*.

branchial members are converted into two pairs of simple three-jointed tufted feet. The hairy tufts are appended only to the last joint. The terminal segment of the six-jointed tail (fig. *h*) is expanded into a large quadrangular surface, the inferior margin of which is fringed with six pairs of long slender filaments.

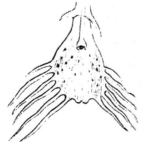

Fig. *h*.

INTRODUCTION.

But one of the most remarkable peculiarities of this state of the animal is the existence of an excessively long filament extending from above the eyes and in front of the corselet: this is rough with minute spines, and appears, as Mr. Couch says, to be hollow. Two similar filaments, equally long, are attached to the posterior part of the corselet above the tail. From the repeated and careful observations of Mr. Couch there can be no doubt of their correctness, for he not only bred them repeatedly in filtered water, but succeeded in artificially extracting some from the ova.*

The metamorphosis in the Macroura generally is less strongly marked than in those forms to which we have hitherto referred. Of these, the lobster, *Homarus,* the spiny lobster, *Palinurus,* the prawn, *Palemon,* and the shrimp, *Crangon,* have been more or less fully observed. Mr. Brightwell did not consider the changes which he observed in the lobster such as to warrant the application of the term metamorphosis; but even Mr. Rathke himself, whose researches in the river species have offered the strongest arguments to the opponents of this view, in his subsequent Memoir, adduces this, amongst other species, as an attestation of the truth of Mr. Thompson's theory. Mr. R. Couch's figure (fig. *i*) of the young lobster on its exit from the egg does not differ materially from that

Fig. *i.*

---

* The larva of *Galathea* is figured at p. 203, in illustration of Mr. R. Couch's description at the previous page.

of *Galathea* and *Palinurus*, excepting that on the superior rings of the tail in the latter are situated four pairs of appendages (fig. *j*). Upon this point Mr. Couch has the following sensible remarks.

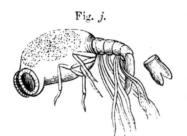

Fig. *j*.

" There is not certainly the same difference of configuration between the young and adult condition of these (the Macroura), as is found to be the case with the short-tailed crabs, simply from the circumstance of the tail being extended in both states, and the claws also show a nearer approach to each other. But this similarity is more apparent than real, for the physiological difference is nearly as wide in one case as in the other. . . . . The form of the shield and the body generally, the sessile character of the eyes, and the long and slender filaments on the tail in *Palinurus*, undergo an entire change in the transformation. The shield and body become more depressed and elongated, the eyes become elevated on stout footstalks," &c. The sessile character of the eyes in the early stage of all the *Podophthalma* hitherto examined is a very remarkable and important character.

The changes in the smaller decapod Macroura, represented by the genus *Palemon*, were first examined by Mr. Thompson, and formed the subject of a second paper read before the Royal Society in 1836. This paper, as well as that on *Carcinus* before referred to, appears not to have been known to the late Capt. du Cane, who having amused the hours of a long illness by a number of interesting investigations on subjects of Natural History, communicated two papers to the Annals of Natural History, on the Metamorphosis of Crustacea. To one of these, on *Carcinus Mænas*, I have already alluded; the other * contains a brief account of the transformation and development of the ditch prawn, *Palemon varians*, in four stages, accompanied by excellent figures; and a still more slight one of the common shrimp, *Crangon vulgaris*, in its first stage only. I give that author's figure of the first stage of the prawn (fig. *k*), in which the locomotive organs are probably the homologues of the foot-jaws, and the rudiments of some of the true feet appear under the cephalo-thorax. The eyes are wholly sessile; there is not the slightest appearance of abdominal members; and the simple spatulate form of the tail is remarkably different from the highly developed and complicated structure of that organ in the adult.

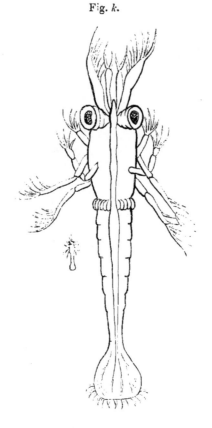

Fig. *k*.

The two following figures exhibit two successive states of the young animal, gradually approaching more and more to the adult condition. In Fig *l*, is seen one of the serratures of the

---

* Ann. Nat. Hist. vol. ii. p. 178, pl. vi. and vii.

INTRODUCTION.

Fig. *l.*

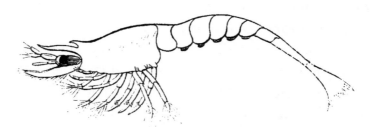

carapace, the rostrum is produced, the important change in the eyes from the sessile to the pedunculate form has now taken place, the true feet have become evolved, developed, and rudimentary abdominal members are perceived, and in fig. *m*, all these developments are far more advanced, and the animal has nearly approached its final state.

Fig. *m.*

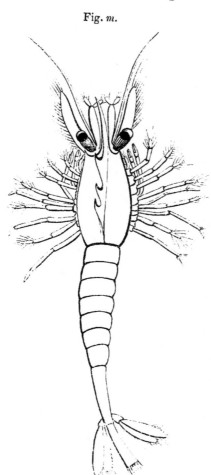

Such is, as far as it appeared to me necessary to detail it, the state of our present knowledge of this very interesting phase in the economy of this class of animals. I have entered more into detail in the history of the discovery, in order to do justice to those whose original and independent observations led them to break through the trammels of preconceived notions, and, notwithstanding much opposition and some misrepresentation, persevered in prosecuting the investigation until the truth of the doctrine has been universally received.*

It has not been my object, in the present Introduction, to enter into the details of the anatomy and physiology of the class of animals of which it treats. It has been considered sufficient for my present purpose to offer a very slight sketch of the principal organs and their functions,

---

* During the passage of part of this introduction through the press, I received a communication from my friend Mr. Couch, containing some new observations on the development of the lobster. From these observations, and others made on Caprella and other forms, Mr. Couch comes to the following conclusions, which are strongly confirmatory of the doctrine of arrested development, and are, in that point of view, very interesting. The original paper was read at the Royal Cornwall Polytechnic Society.

"So far as my observation has extended, it appears probable that the metamorphosis of the young in their progress to adult growth is not universal in all Crustaceans; but, on the contrary, that the families in which the eyes are always sessile in their adult growth, and which do not exuviate or voluntarily throw off their limbs, are in the habit of producing their young perfectly formed; and an opportunity that has occurred to me of observing the process of early development in the common lobster will tend to establish the existence of a law of Nature as applicable not only to it, but probably also to all the genera of this extensive family or class—that is, the long-tailed crustacea—which law is, that the greatest extent of metamorphosis is in those genera which are of the highest rank in the series—that is, the short-tailed, or crabs—that, even at their birth, the long-tailed genera—as the lobster—approach more closely to the ultimate form of the parent; and—what is still more extraordinary than all beside—that *so long as the lobster* in particular, *retains the eyes sessile*, the progress of *development and growth* is conformed to what is the perpetual *mode* of *growth* of the *permanently sessile-eyed races;* and it is only when the crust has become fully extended and hardened, and thus the exuviation is rendered necessary, that the *eyes become elevated* on footstalks, and the adult *form and habit* are completely established."

with reference, on the one hand, to the characters which are given of the different genera and species, in the body of the work, and, on the other, to their habits and mode of life. For those who seek for further information, I beg to refer to the excellent digest contained in Professor Rymer Jones's " Outline of the Animal Kingdom," to Professor Owen's admirable lectures on the Invertebrata, to Dr. Milne Edwards's article CRUSTACEA in Dr. Todd's "Cyclopædia of Anatomy and Physiology," and above all, to the great general work of the same author on the natural history of this class of animals.* In the introductory portion of that invaluable book, and in the general description of the different groups contained in the body of the work, will be found an immense fund of information, great part of which is original and based upon the actual dissection and observation of that distinguished naturalist, and of his no less talented friend and coadjutor, Mons. Audouin. My obligations to this unrivalled monograph will appear in every page of this little work, and demand my warmest acknowledgments.

* Histoire Naturelle des Crustacés, tom. iii. Paris, 1834.

INDEX.

[The systematic names, including the Latin synonyms, are printed in *Italics*.]

A.

Achæus Cranchii, 10.
Alauna rostrata, 330.
Albunea dentata, 159.
Alpheus Caramote, 318.
„ *ruber*, 271.
„ *sivado*, 312.
„ *Spinus*, 284.
Astacus Bamfficus, 208.
„ *Bernhardus*, 171.
„ *Crangon*, 256.
„ *fluviatilis*, 237.
„ *Homarus*, 213.
„ *marinus*, 242.
„ *nitescens*, 261.
„ *Norvegicus*, 251.
„ *serratus*, 302.
„ *squamifer*, 197.
„ *Squilla*, 305.
„ *strigosus*, 200.
„ *stellatus*, 223.
„ *subterraneus*, 217.
Atelecyclus heterodon, 153.
„ *septemdentatus*, 153.
Athanas nitescens, 261.
Axius stirynchus, 228.

B.

Blastus tetraodon, 22.
Bodotria arenosa, 333.

C.

Callianassa subterranea, 217.
Calocaris Macandreæ, 233.
Cancellus marinus, 135.
Cancer angulatus, 130.
„ *araneus*, 31.
„ *asper*, 46.
„ *Astacus*, 237.
„ *Bamfficus*, 208.
„ *Bernhardus*, 171.
„ *biaculeatus*, 27.
„ *Bufo*, 31.
„ *Cassivelaunus*, 159.
„ *corrugatus*, 94.

Cancer denticulatus, 72.
„ *Depurator*, 101.
„ *digitatus*, 351.
„ *Dorsettensis*, 13.
„ *Dromia*, 369.
„ *floridus*, 51.
„ *fluviatilis*, 237.
„ *Gammarus*, 242.
„ *hirtellus*, 68.
„ *horridus*, 165.
„ *hydrophilus*, 54.
„ *inciso-serratus*, 59.
„ *incisus*, 51.
„ *latipes*, 85.
„ *longicornis*, 193.
„ *Maia*, 39, 165.
„ *minutus*, 135.
„ *Mænas*, 76.
„ *Norvegicus*, 251.
„ *Pagurus*, 59.
„ *Phalangium*, 2.
„ *Pisum*, 121.
„ *Platycheles*, 190.
„ *puber*, 90.
„ *rostratus*, 2.
„ *Scorpio*, 13.
„ *Spinus*, 284.
„ *Squilla*, 202, 305.
„ *Squinado*, 39.
„ *strigosus*, 200.
„ *tetraodon*, 22.
„ *tuberosus*, 141.
„ *tumefactus*, 145.
„ *velutinus*, 90.
Carcinus Mænas, 76.
Corystes Cassivelaunus, 159.
„ *dentatus*, 159.
Crab, angular, 130.
„ circular, 153.
„ floating, 135.
„ great, 59.
„ harbour, 77.
„ masked, 159.
„ shore, 77.
Crangon bispinosus, 268.
„ *cataphractus*, 261.
„ *fasciatus*, 259.
„ *sculptus*, 263.

Crangon spinosus, 261.
„ *trispinosus*, 265.
„ *vulgaris*, 256.
Cray-fish, common, 237.
Cuma Audouinii, 328.
„ *Edwardsii*, 326.
„ *trispinosa*, 329.
Cynthia Flemingii, 379.

D.

Dromia Rumphii, 369.
„ *vulgaris*, 369.

E.

Ebalia Bryerii, 145.
„ *Cranchii*, 148.
„ *Pennantii*, 141.
Eurynome aspera, 46.
„ *spinosa*, 46.

G.

Galathea Bamffia, 208.
Galathea, Embleton's, 204.
Galathea nexa, 204.
„ *rugosa*, 208.
Galathea, scaly, 197.
Galathea spinigera, 200.
Galathea, spinous, 200.
Galathea squamifera, 197.
„ *strigosa*, 200.
„ *deltura*, 225.
Gebia stellata, 223.
Gelasimus Bellii, 130.
Gonoplax angulatus, 130.
„ *bispinosus*, 130.
„ *rhomboides*, 130.
Grapsus cinereus, 135.
„ *minutus*, 135.
„ *Testudinum*, 135.

H.

Hermit-crab, blue-banded, 375.
„ common, 171.
„ Prideaux's, 175.
„ rough-clawed, 186.
„ smooth, 184.
Hippa septemdentata, 153.
Hippolyte Cranchii, 288.
Hippolyte, Cranch's, 288.
Hippolyte Moorii, 292.
„ *Pandaliformis*, 294.
„ *Prideauxiana*, 292.
Hippolyte, Prideaux's, 292.
Hippolyte Sowerbæi, 284.
Hippolyte, Sowerby's, 284.
Hippolyte Spinus, 284.
„ *Thompsoni*, 290.
Hippolyte, Thompson's, 290.

Hippolyte varians, 286.
Hippolyte, varying, 286.
Homarus vulgaris, 242.
Hyas araneus, 31.
„ *coarctatus*, 35.

Inachus araneus, 31.
„ *Dorsettensis*, 13.
„ *Dorynchus*, 16.
„ *leptochirus*, 18.
„ *Phalangium*, 2.
„ *Scorpio*, 13.

L.

Leptopodia tenuirostris, 6.
Lithodes Maia, 165.
Lobster, common, 242.
„ Norway, 251.

M.

Macropodia longirostris, 6.
„ *Phalangium*, 2.
„ *Scorpio*, 13.
„ *tenuirostris*, 6.
Macropus longirostris, 6.
„ *Phalangium*, 2.
Maia aranea, 31.
„ *Squinado*, 39.
„ *tetraodon*, 22.
Mantis digitatus, 351.
Munida, long-armed, 208.
Munida Rondeletii, 208.
„ *rugosa*, 208.
Mysis Chamæleon, 336.
„ *Griffithsiæ*, 342.
„ *Leachii*, 336.
„ *rostratus*, 342.
„ *spinulosus*, 336.
„ *vulgaris*, 339.

N.

Nautilograpsus minutus, 135.
Nephrops Norvegicus, 251.
Nika canaliculata, 275.
„ *Couchii*, 278.
„ *edulis*, 275.

O.

Ocypoda angulata, 130.

P.

Pagurus Bernhardus, 171.
„ *Cuanensis*, 178.

INDEX.

Pagurus Dillwynii, 377.
," *fasciatus*, 375.
," *Forbesii*, 186.
," *Hyndmanni*, 182.
," *lævis*, 184.
," *Prideauxii*, 175.
," *streblonyx*, 171.
," *Thompsoni*, 372.
," *ulidianus*, 180.
Palæmon Leachi, 307.
," *nitescens*, 261.
," *serratus*, 302.
," *Squilla*, 302, 305.
," *varians*, 309.
Palinurus Homarus, 213.
," *quadricornis*, 213.
," *vulgaris*, 213.
Pandalus annulicornis, 297.
Pasiphæa brevirostris, 312.
," *Savignii*, 312.
," *Sivado*, 312.
Pea-crab, Pinna, 126.
," common, 121.
Penæus Caramote, 318.
," *trisulcatus*, 318.
Planes Linnæana, 135.
Platycarcinus Pagurus, 159.
Pilumnus hirtellus, 68.
Pinnotheres Cranchii, 121.
," *Latreillii*, 121.
," *Montagui*, 126.
," *Pinnæ*, 126.
," *Pisum*, 121.
," *varians*, 121.
," *Veterum*, 126.
Pirimela denticulata, 72.
Pisa biaculeata, 27.
," *Gibbsii*, 27.
," *tetraodon*, 22.
Pisidia longicornis, 193.
Platyonicus latipes, 85.
Polybius Henslowii, 116.
Pontophilus bispinosus, 268.
," *spinosus*, 261.
," *trispinosus*, 265.
Porcelain-crab, hairy, 190.
," minute, 193.
Porcellana Leachii, 193.
," *longicornis*, 193.
," *platycheles*, 190.
Portumnus variegatus, 85.
Portunus arcuatus, 97.
," *corrugatus*, 94.
," *Dalyellii*, 361.
," *Depurator*, 101.
," *emarginatus*, 97.
," *holsatus*, 109.
," *infractus*, 361.
," *lividus*, 109.

Portunus, longipes, 361.
," *maculatus*, 112.
," *marmoreus*, 105.
," *Mœnas*, 76.
," *plicatus*, 101.
," *puber*, 90.
," *pusillus*, 112.
," *Rondeletii*, 97.
Prawn, 303.
Processa canaliculata, 275.
," *edulis*, 275.

S.

Shrimp, banded, 259.
," common, 256.
," sculptured, 263.
," spinous, 261.
," three-spined, 265.
," two-spined, 268.
Spider-crab, Cranch's, 11.
," four-horned, 22.
," Gibbs's, 27.
," long-legged, 2.
," Scorpion, 13.
," slender, 6.
," slender-legged, 19.
," spinous, 39.
Spiny-lobster, common, 213.
Squilla Desmarestii, 354.
," *Mantis*, 351.
Stenorynchus longirostris, 6.
," *Phalangium*, 2.
," *tenuirostris*, 6.
Stone-crab, northern, 165.
Swimming-crab, arched, 97.
," cleanser, 101.
," dwarf, 112.
," Henslow's, 116.
," livid, 109.
," long-legged, 361.
," marbled, 105.
," velvet, 90.
," wrinkled, 94.

T.

Themisto brevispinosa, 384.
," *longispinosa*, 381.
Thia polita, 365.
Thysanopoda Couchii, 346.

X.

Xantho florida, 51.
," *inciso-serrata*, 51.
," *rivulosa*, 54.
," *tuberculata*, 359.

BRITISH CRUSTACEA.

DECAPODA.   *LEPTOPODIADÆ.*
  BRACHYURA.

GENUS STENORYNCHUS, Lamarck.

Cancer,	Linn. Penn. Herbet.
Inachus,	Fabr.
Maia,	Bosc.
Macropus,	Latr.
Macropodia,	Leach.
Stenorynchus,	Lamk. Edwards.

Generic character. *External antennæ* setaceous, the basal joint narrow, the second * inserted close to the side of the rostrum, very short; the third, three times as long as the former. *External pedipalps* narrow, the second joint considerably produced internally at its apex; the third joint oval. *Anterior feet* shorter, and much larger (in the male) than the succeeding ones; equal; the hand somewhat ventricose; the fingers slightly inflected. The remaining pairs very long and slender, diminishing in length from the second to the fifth; the nails of the second and third pairs long, slender, and curved only at the apex; those of the fourth and fifth shorter, curved at the base and somewhat falciform. *Eyes* not retractile, larger than their peduncles, oval, pointed at the apex and setigerous. *Carapace* triangular; *rostrum* taper and bifid. *Abdomen* six-jointed, the terminal portion being formed by the union of the sixth and seventh joints.

   * Leach calls this the first joint of the antennæ, as he does not reckon the basal joint, which is fixed, and, as it were, soldered to the parts contiguous, as in most of the higher forms of Crustacea.

DECAPODA.
BRACHYURA.

LEPTOPODIADÆ.

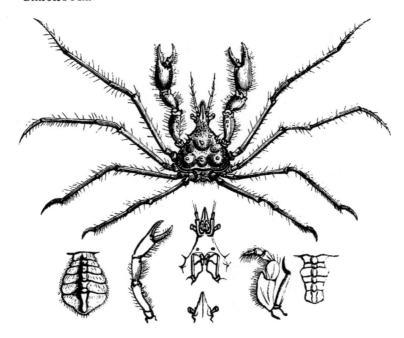

LONG-LEGGED SPIDER-CRAB.

Stenorynchus Phalangium.

Specific character.—Rostrum shorter than the peduncle of the antennæ; a single minute spine on the epistome, close to the auditory cavity; no spine behind the base of the antennæ; arms slightly scabrous, without spines.

Cancer rostratus,		Lin. Faun. Suec. Herbst. t. xvi. f. 90.
„	*Phalangium,*	Penn. IV. t. ix. f. xvii.
Inachus	„	Fabr. Supp. p. 358.
Macropus	„	Latr. Hist. Nat. Crust. VI., p. 110.
Macropodia	„	Leach, Tr. Linn. Soc., XI., p. 331, Malac. Brit. t. xxiii. f. 6.
Stenorynchus	„	Edw. Hist. Crust. I. p. 279.

The general form of the carapace in this species is that of an acute angled triangle, rounded at the posterior angles. It has several rather prominent spines; one on each hepatic region, forming, with one on the gastric region, an equilateral triangle; there are two small ones on each branchial region, and one, the largest of all, on the cardiac; there are also one or two smaller ones near the latero-anterior margin. The rostrum is of moderate length, scarcely reaching to the middle of the third joint of the peduncle of the antennæ; it has a groove through its whole length, reaching to the back of the orbit. The external antennæ are long and setaceous, and furnished with several long hairs; the basal joint is narrow, entirely immoveable, and continuous with the epistome; the moveable part of the peduncle consists of two joints, of which the second is three times as long as the first. The internal antennæ are lodged in fossæ, which are separated from each other by a ridge, which is incomplete at the middle. The eyes are oval, larger than their peduncles, and pointed at the apex, where there is a small bristle.* The orbits are round, and there is a prominent ridge over the upper margin. The epistome, or that portion of the shell between the mouth and the base of the antennæ, has a very minute tubercle, just in front of the organ of hearing, but none at the base of the antennæ, as in *St. tenuirostris*.† The first pair of legs in the male are about twice as long as the body; the arm has a line of minute tubercles on the outer, and another on the inferior surface, which parts are also hairy; but there are no spines on its inner margin, as in *St. tenuirostris:* the wrist is similarly furnished: the hand is somewhat ventricose; it is hairy both on the outer and inner margin; the fingers are slightly inflected; the moveable one is furnished with a tubercle near its base,

---

* This curious appendage I have never seen mentioned as appertaining to this genus.

† This second tubercle is also found in a Mediterranean species *St. Ægyptius.*

and there is a corresponding excavation in the other. In the female these feet are altogether much smaller than in the male. The remaining pairs are very slender and filiform; the second pair is three times and a half the length of the post-rostral part of the body, and they diminish regularly to the last pair: the claws of the second and third pairs are slender, and slightly curved towards the extremity; those of the fourth and fifth are shorter, and somewhat falcate, being curved more abruptly near the base. The abdomen in both sexes has six joints, the sixth and seventh being united into one piece. That of the male is broadest at the base, and again at the union of the third and fourth joints, and terminates in an obtuse triangle: each joint is furnished with a tubercle. The abdomen of the female is very broad, and much curved: the tubercles pass into a continuous obtuse carina on the three or four last joints.

These characters belong for the most part to both the species, excepting where the contrary has been stated; the specific difference with those exceptions being rather in the degree of development than in the existence, or non-existence of parts.

This is one of the most common species of the Triangular Crabs, being found in considerable numbers on most parts of the coast. I have obtained it from Wales, the coast of Cornwall, Devonshire, Dorsetshire, and Sussex, from Scarborough, and from Orkney. It is also not uncommon on the coast of Ireland. Dr. Leach mentions its being particularly common at the mouths of rivers, and in estuaries; but I have found it in very different localities, having often dredged it in deep water, and taken it in crab and lobster pots. Mr. Hailstone states that " it is very common at Hastings, both among the rocks on the shore, and in deep water, and is occasionally caught in the trawl-net in vast numbers. Of sixty-eight specimens brought up at once, the proportion of males to females was as two to one." Like all the species of the family it is slow, sluggish, and timid. It generally has small fuci growing on it, especially on the legs; and I have sometimes seen the body completely covered and concealed by a mass of sponge. When taken it moves with very little energy, and speedily dies after being taken out of the water. Its slow and sluggish habits render it an easy prey to many fishes; Mr. W. Thompson says, "On opening a thornback, *Raia clavata*, about twenty inches in length, I found the stomach entirely filled with *Macropodia Phalangium*."

It deposits its spawn during the early spring months.

DECAPODA.
BRACHYURA.

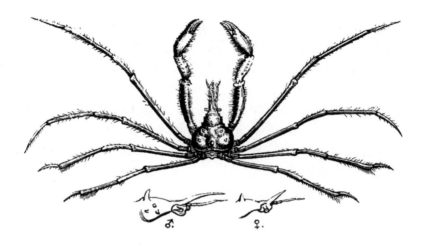

SLENDER SPIDER-CRAB.

Stenorynchus tenuirostris.

Specific character.—Rostrum longer than the peduncle of the external antennæ, its two portions being in contact throughout their whole length; two minute spines on the epistome, one close to the organ of hearing and another at the base of the external antennæ; arms spinulose at the inner margin.

Leptopodia tenuirostris,	LEACH, Edinb. Enc. VII., p. 431.
Macropus longirostris?	LATR. Hist. Nat. des Crust. VIII., p. 110.
Macropodia „	RISSO. Hist. Nat. de l'Eur. Merid. V., p. 27.
„ tenuirostris,	LEACH, Trans. Linn. Soc. XI. p. 331. Malac. Brit. t. XXIII. f. 1–5.
Stenorynchus longirostris?	EDW. Hist. Crust. I. p. 280. COUCH, Cornish Fauna, p. 64.

THIS elegant species may be readily distinguished from the former by the long attenuated rostrum, by the existence of a small spine on the epistome, immediately behind the basal joint of the external antennæ, and by a series of minute spines on the inner part of the arm. The body is altogether more elongated, and the spines more acute; but, in other respects, the characters are nearly the same.

"I first observed this species," says Dr. Leach, "amongst some Crustacea collected at Torquay, in Southern Devon, by Hooker; and have since found it a very common inhabitant of all the deep water off the coast of that country, especially in the Sound of Plymouth." Mr. Couch states it to be very common in Cornwall, at the depth of from two to twenty fathoms; and Mr. Embleton includes it in his list of the Crustacea of Berwickshire and North Durham. It does not appear to have been taken in Ireland. I have taken it in prawn pots at Bognor, and by dredging in Studland Bay in Dorsetshire.

I have appended a note of doubt to the synonyms of the Mediterranean species, *Macropus longirostris*, Latr., hitherto considered as identical with this, as I am much inclined to believe they may be distinct. I am led to this supposition by a careful examination of specimens of my own collection on our coast, with some which I had received from Sicily, and from the Bay of Naples, and I find that on all those brought from the Mediterranean, the body is proportionally longer; the rostrum also longer and more slender, reaching very much beyond the peduncle of the antennæ. By measurement I find that, in the Mediterranean specimens, the length of the carapace, including the rostrum, is to its breadth, at the widest part, as five to two; whereas, in the British, it is not quite twice as long as broad. The two portions of the rostrum in the former are a little separated throughout almost their whole length, and each is perfectly round; whereas, in the British specimens, they are entirely in contact, and flattened above and beneath. There are a few other differences principally proportional, but these are the most considerable. These may be mere accidental variations, but I think it not improbable that they indicate a specific distinction.

DECAPODA.
BRACHYURA.

LEPTOPODIADÆ.

GENUS ACHÆUS, Leach.

ACHÆUS,   Leach, Latr. Edwards.

Generic character. External antennæ remote, setaceous, the first articulation united to the front, and extending beyond the inner canthus of the orbits; the second articulation inserted at the side of the rostrum, and entirely exposed from above, and, with the third, much thicker than the subsequent ones. *External pedipalps*, with the second articulation much longer than broad, and produced at the interior and anterior angles, the third subtriangular with the angles rounded. *The first pair of feet* (in the female) short, rather slender; the second and third pairs having the terminal joint long and styliform; that of the fourth and fifth compressed, abruptly curved, and falciform. *Carapace* somewhat triangular, slightly spinous, the branchial regions elevated and swollen. *Rostrum* extremely small, bifid. *Eyes* not retractile, placed on long footstalks of equal size, and furnished with a single tubercle on the fore-part. *Abdomen* six-jointed in both sexes.

This genus, of which one species only is at present known, is considered by Dr. Leach as intermediate between *Inachus* and *Leptopodia* [*Macropodia*], and by Milne Edwards it is placed between *Stenorynchus* (*Macropodia* Le.) and *Camposcia*. Its relation to *Eurypodius* is also probable from the character of the feet, whilst the structure of the eyes and some other points appear to indicate an approach to some of the *Maiadæ*.

DECAPODA.
BRACHYURA.

LEPTOPODIADÆ.

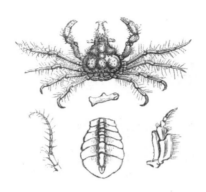

CRANCH'S SPIDER-CRAB.

Achæus Cranchii. Leach.

Specific character.—" Carapace, with two tubercles in the median line, and with two elevated lines between the eyes."—Leach.

Achæus Cranchii,   Leach, Malac. Brit. XXII. C. Ed. 2. Latr. Reg. Anim. IV. p. 64.   Edw. Hist. Crust. I. p. 281.

The carapace of this interesting species, is triangular, contracted behind the orbits, then enlarged into a prominent point or tubercle, then again contracted, and finally enlarged and rounded at the sides of the branchial regions. Two conspicuous elevations, or tubercles, occur on the median line, with an inconspicuous one between them; and the branchial regions are elevated and rounded. The rostrum is extremely small and bifid, as broad as it is long. The orbits are small and open above, and the eyes exposed almost to the insertion of the peduncles, which are long, cylindrical, furnished with a small rounded tubercle on the anterior part, about the middle of its length, and standing directly outwards; not retractile. The antennæ and the feet are very hairy. The hands are carinated longitudinally. The epistome is quadrate. The abdomen in the female (and, according to Dr. Milne Edwards, in the male also,) is six-jointed. In the former it is oval, expanded towards the posterior part, and carinated through its whole length. The carapace is about six lines in length.

Colour, pale reddish brown.

Of the occurrence of this beautiful little species on our coasts, we have, I believe, only two recorded instances. In the " Malacostraca Podophthalma Britanniæ," Dr. Leach first made it known as having been discovered by Mr. Cranch in dredging off Falmouth. This single specimen, a female, is now in the British Museum. The second example is thus stated by Mr. W. Thompson in his catalogue of the Crustacea of Ireland. " In the collection of Crustacea formed by Mr. J. V. Thompson, and now in the possession of the Royal College of Surgeons, Dublin, is a native specimen of this crab, which we may presume was obtained on the Southern coast." This is the sum of the information we have respecting this species as indigenous to this country. Dr. Milne Edwards gives as its habitat on the French coast, " l'embouchure de la Rance, près Saint-Malo." Of its habits nothing whatever is recorded, beyond the remark of Dr. Edwards, that it lives amongst sea-weeds and on oyster-beds.

DECAPODA.
BRACHYURA.

LEPTOPODIADÆ.

GENUS INACHUS, Fabr.

CANCER,	Pennant, Herbst.
INACHUS,	Fabr. Leach, Latr. Edw.
MACROPUS,	Latr.
MAIA,	Bosc.

Generic character.—*External antennæ* not more than one-fifth of the length of the body; the basal joint forming the inferior margin of the orbit; the second inserted by the side of the rostrum. *External pedipalps* with the second joint much produced internally; the third joint elongate, somewhat triangular, the anterior and inner angle truncate at the insertion of the palp, which is three-jointed. The *anterior legs*, in the male, twice as long as the body, the arms and the hands subovate, the fingers inflected. The remaining pairs very long, diminishing in length from the second to the fifth; second pair larger than the succeeding ones; the terminal joint long and slightly curved. *Carapace* subtriangular, nearly as broad as long, the rostrum short and bifid. *Eyes* on short footstalks, retractile or capable of being bent backwards and lodged in the posterior part of the orbit. *Abdomen* in both sexes, six-jointed and carinated.

DECAPODA.   *LEPTOPODIADÆ.*
BRACHYURA.

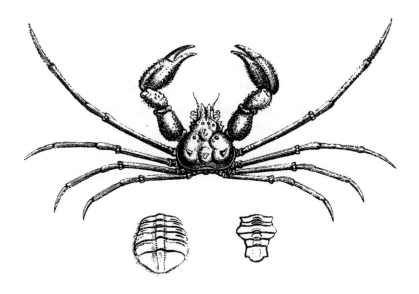

SCORPION SPIDER-CRAB.

Inachus Dorsettensis. Leach.

Specific character.— Rostrum very short, emarginate: the gastric region furnished with four small tubercles ranged in a line transversely, and a larger one behind them.

Cancer Dorsettensis,	PENN. Brit. Zool. IV. t. x. f. 1. p. 12.
„ *Scorpio*,	FABR. Ent. Syst. II. p. 462. HERBST. I. p. 237. No. 130.
Macropus „	LATR. Hist. Nat. Crust. VI. p. 109.
Inachus „	FABR. Suppl. 358. DESM. Cons. t. xxiv. f. 1. EDW. Hist. Crust. I. p. 288. COUCH, Cornish Fauna, p. 65.
„ *Dorsettensis*,	LEACH, Edinb. Encycl. Malac. Brit. t. xxii. f. 1–6.

I HAVE found it necessary to restore to this species the original specific name given to it by Pennant, who first described it from specimens in the Portland Cabinet, taken at Weymouth, from which locality he designated it *Cancer Dorsettensis*. His work was published in 1777; and the Entomologia Systematica of Fabricius, in which it first received the name of *C. Scorpio*, not until 1793. The Fabrician name has recently been adopted by Dr. Milne Edwards, as it had previously been by Desmarest, probably from some objection to the local origin of the former name; this, however, is quite admissible in the present instance, as indicating the locality in which it was first discovered. At all events, it is not more objectionable than the other.

The carapace of this species is triangular, rounded posteriorly, and ventricose. The rostrum is very short and bifid; the orbits oval, so that the eyes, which are attached by their peduncles to the anterior portion of the orbit, can be laid backwards into the posterior portion of that cavity; a character which belongs to most of the genera of the triangular or oxyrynchian families. The eyes are protected by a spine on the anterior, and a stronger one on the posterior margin of the orbits, of which the upper margin is also raised, and the inferior, formed by the basal joint of the antennæ, slightly tuberculated. The external antennæ are short; the moveable portion not much exceeding twice the length of the rostrum. There are four small tubercles on the anterior part of the carapace arranged transversely, and one much larger behind them, on the centre of the gastric region; there are two tubercles on each branchial region, one at the anterior part and another rather larger on the centre; there is also a conspicuous one on the cardiac region. The external pedipalps are elongate, the second joint being much produced anteriorly at the inner angle; and the third, which is somewhat triangular, has the inner and anterior angle truncated, for the articulation of the terminal portion, which consists of three joints. The anterior pair of legs in the male are thick and long, the joints of a somewhat oval form, and the fingers considerably incurved. Those of the female are very small. The remaining feet are very long and slender, the second pair being considerably more than three times the length of the body, including the rostrum. They are also much larger than the succeeding one, which diminish in length and thickness to the last. The abdomen of the male is rather short and broad, the widest part being at the union of the third and fourth joints; that of the female is remarkably broad. In both sexes it is tuberculo-carinated.

It would appear that this species is more widely distributed than had been supposed. Dr. Leach states that it is very plentiful on the coast of Devon; we have seen that Pennant's specimens were from Weymouth; and I obtained it in Studland Bay, Dorsetshire, and at Hastings. Mr. Couch states that in Cornwall it is commonly taken in crab pots, within a few miles of the shore, at all depths; and Mr. Eyton informs me that it is found on the oyster-beds at Rhoscolyn, near Holyhead. In Ireland it has been found in many places; in the Harbour of Cove, by Mr. J. V. Thompson. "It is pretty commonly taken," says Mr. W. Thompson, "in the loughs of Strangford and Belfast, and on the western coast.—Mr. Ball," adds Mr. Thompson, "finds it in Dublin Bay." It is also recorded that Captain Beechey, R.N., brought up a specimen of this species alive in the dredge from a depth of one hundred and forty fathoms, in the Mull of Galloway. Its habitat extends far north, Fabricius having found it in the Norwegian Seas.

DECAPODA.
BRACHYURA.
LEPTOPODIADÆ.

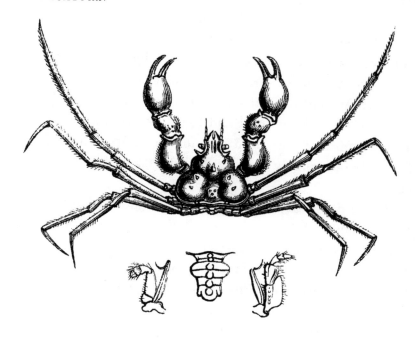

Inachus Dorynchus. Leach.

Specific character.—Rostrum bifid, extending beyond the third joint of the peduncle of the antennæ ; gastric region with three spines, two anterior, and the third much longer, forming a triangle. Second pair of legs not more than three times the total length of the body.

Inachus Dorynchus, Leach, Edinb. Enc. art. Crust. p. 431. Id. Malac. Brit. t. xxii. f. 7-8. Edw. Hist. Crust. p. 288. Couch, Cornish Fauna, p. 65.

The general form of this species is very similar to the former, but it is less globose. The carapace is triangular, longer than it is broad. The rostrum is short, somewhat hastiform, and in most slightly bifid; although in some specimens the division is more considerable. The antennæ, the eyes, and orbits, as well as the external pedipalps, are very similar to those of the former species. The gastric region of the carapace has three spines, two small ones distant, and another much stronger placed farther back on the median line, and, with the others, forming a triangle. There are two tubercles on each hepatic region, placed as in the former species; and the cardiac region, instead of a spine, has only an elevation, on which are three very small tubercles. The sides of the shell are destitute of tubercles. The hands are smooth. In other respects this species resembles the former.

The present species of *Inachus* was discovered by Dr. Leach, as he informs us, " whilst cleaning a parcel of *I. Dorsettensis* from the Salcombe estuary for examination." Mr. Couch states that it is commonly found in crab-pots in Cornwall. Mr. Hailstone found it at Hastings, where I have also obtained it. I have taken it by the dredge in Studland Bay, Dorsetshire, and at Bognor I found several small specimens amongst the refuse of prawn and lobster pots. These were of a lighter colour than most which I have observed from other localities, but this may have arisen from their being young. In Mr. Embleton's list of the Crustacea of Berwickshire and North Durham, it is stated to occur not uncommonly in Berwick and Embleton Bays. It is found on the coast of Ireland, though rarely, having been taken by Dr. Drummond in Belfast Bay.

This species, like all the others of the family, is very liable to be covered with small fuci and sponges; hence, as Dr. Leach has observed, in all probability arose its having been for so long a time undiscovered, having doubtless been passed over as *I. Dorsettensis;* it does not, however, at present appear to be so generally distributed as that species.

DECAPODA.
BRACHYURA.

LEPTOPODIADÆ.

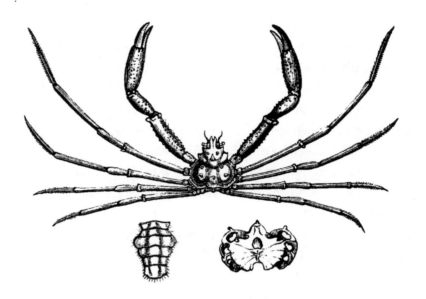

SLENDER-LEGGED SPIDER-CRAB.

Inachus leptochirus. Leach.

Specific character.—Feet slender, anterior pair in the male extending beyond the penultimate joint of the second pair. Rostrum hastiform. Sternum in the male with a round polished tubercle.

Inachus leptochirus,   Leach, Malac. Brit. t. xxii. B.
                          (errore *leptorinchus*) Edw. Hist. Crust. I. p. 289.

The carapace in this species considerably resembles that of *I. Dorynchus*. It is triangular, considerably longer than it is broad, much narrowed forwards; the rostrum hastiform, bifid at the extremity, and with a slight groove extending from thence backwards between the eyes. There is a strong spine on the gastric region, a very small tubercle on each hepatic, a spine on the latero-anterior margin, two on each branchial region, the posterior being the larger, and one on the genital region in a straight line between the two larger ones on the branchial. The feet are all very long and slender. The hands in the adult male are considerably longer than the carapace; the fingers curved. The second pair of feet are three times the length of the carapace. On the sternum, immediately in front of the apex of the abdomen, when in its usual position applied against the thorax, is a round or oval prominent and polished tubercle, of a greyish-white colour.

In the adult state this is considerably the largest of the British species of *Inachus*. It is also then readily distinguished from the others, by the general form, as well as by the extraordinary length of all the legs, and especially by the form and length of the first pair. But in the younger state all these characters are much less conspicuous, and it might almost be mistaken for *I. Dorynchus*, but for the remarkable character of the round polished tubercle on the thorax, which somewhat resembles the half of a pearl. This is peculiar to the male, and cannot fail to strike us as offering a very obvious mark of relation to the Mediterranean species *I. Thoracicus*, on the thorax of which there is a very curious development of a similar hard shelly substance, in the form of a broad, three-lobed plate. This formation is peculiar to the genus *Inachus*, and, as far as it is at present known, to the two species in question.

The *Inachus leptochirus* is extremely rare. It was discovered by the ill-fated Mr. Cranch on the western coast of Devon, or Cornwall, and was afterwards taken by Mr. Prideaux from a crab-pot in Bigbury Bay. In Mr. W. Thompson's "Additions to the Fauna of Ireland," is mentioned "a specimen dredged in Clifton Bay, Connemara, by Mr. Forbes and Mr. Hall, and another in Belfast Bay by Mr. Patterson." The latter specimen, through the kindness of Mr. Thompson, I have now before me. It is a young male. The same gentleman subsequently states that he had seen specimens from Belfast Bay "in the Ordnance collection." This is the extent of our knowledge of this curious species.

Dr. Milne Edwards has misquoted Leach's specific name as "*Leptorinchus,*" and this error has been copied by Mr. Couch in his " Cornish Fauna."

DECAPODA.
BRACHYURA.

MAIADÆ.

GENUS PISA.

CANCER,	Penn. Herbst, Montagu.
INACHUS,	Fabr. Risso.
MAIA,	Latr. Bosc.
PISA,	Leach, Desmar, Edwards.

Generic character.—*External antennæ* beset with club-shaped nairs; the basal joint longer than broad, extending beyond the inner canthus of the orbit; but concealed above by the strong spine which proceeds from the upper margin of the orbit: second joint of the antennæ rather slender, inserted a little behind and on the outer side of the rostrum. *External pedipalps* very broad, the second joint produced at the inner and anterior angle; the third triangular, very broad at the outer margin; the anterior and inner angle truncate or emarginate. *First pair of feet* in the adult male very large, longer than the second pair; the hand thick and the fingers meeting only at the outer margin of the points which are toothed; those of the female much smaller, the fingers meeting throughout nearly their whole length; shorter than the second pair. The remaining feet moderately long, diminishing regularly from the second to the fifth, cylindrical, the terminal joint curved, pectinato-denticulated beneath, naked at the extremity. *Eyes* scarcely thicker than their peduncles, capable of being reflected in the orbits. *The orbits* oval, directed outwards and downwards; their upper margin with a strong triangular spine directed forwards. *Carapace* triangular, terminating in a strong bifid rostrum, divaricating at the extremity. *Abdomen* seven-jointed in both sexes.

DECAPODA.
BRACHYURA.

MAIADÆ.

FOUR-HORNED SPIDER-CRAB.

Pisa tetraodon. Leach.

Specific character.—Lateral margin with four spines, (exclusive of those above and behind the orbit.) Posterior portion of the carapace rounded, without spines; a small tubercle near the posterior margin.

Cancer tetraodon,		PENN, Brit. Zool. IV. t. viii. f. 2. p. 11.
Maia	,,	Bosc, Hist. Crust. I. 254. LEACH, Edinb. Encycl. VII. p. 395.
Blastus	,,	LEACH, l. c. p. 431.
Pisa	,,	LEACH, Trans. Linn. Soc. XI. p. 328. Id. Encycl. Brit. Supp. I. p. 415. Id. Malac. Brit. t. xx. f. 1–4. EDW. Hist. Crust. I. p. 305. COUCH, Cornish Fauna. p. 65.

THE general form of the body of this species is triangular, produced anteriorly, and with the posterior angles much rounded. The rostrum is large, strong, and prominent, about one-third as long as the remainder of the carapace; it is formed of two strong horns, diverging for about one-third of their length, and slightly deflexed; the lateral margin has four spines, exclusive of a very strong one above the orbit, and a smaller one behind that cavity. There are numerous tubercles on the carapace, several small ones on the gastric region, disposed transversely; one on the centre of the carapace; two considerable ones on each branchial region, one on the centre of the cardiac, and a small one near the posterior margin. The spines above the orbit are triangular, very strong and prominent; directed forwards and a little outwards, and so formed that the eyes can be deflexed within them, so as to be quite concealed from above. The external antennæ are beset at their base with long club-shaped hairs. The anterior pair of feet in the male are exceedingly strong and thick, the hands especially are nearly as broad as they are long. The fingers meet at the points; the outer edge of each being denticulated, and the moveable one has a small round tooth. The arms and wrists have several round tubercles. In the female these feet are very small, and shorter than the second pair, and in the immature male they are very similar to those of the female. The remaining feet are of moderate size and length, the second pair being but little longer than the carapace, and the fifth pair shorter than its breadth. There are a few tubercles, and a few small spines upon the legs, and the nail is furnished beneath with a regular row of sharp spines arranged like the teeth of a comb. The abdomen has seven distinct joints in each sex; that of the male being broadest at the third joint; the sixth is broader than the fifth, and the seventh is triangular. Each joint has a central tubercle.

The abdomen of the female is very large and broad, and has a broad carina. The whole surface of the shell, and the greater part of the limbs, is covered with a close, short, villous coat; and the antennæ, rostrum, and all the tubercles are furnished with tufts of long, curved, club-shaped hairs. Underneath this covering the shell is polished, and minutely punctured. The colour is a dull reddish-brown, becoming bright red by boiling, or by the action of spirit. The general length of the carapace in a full-grown male is two inches three lines, breadth one inch six lines.

The habits of this species, as far as I have had an opportunity of observing them, are curious. They are found concealed under the long hanging fuci which clothe the rocks at some distance from the shore, in which situation I have taken them amongst the Bognor rocks. They congregate in vast numbers at the place I have just mentioned, in the prawn and lobster pots. I have seen, probably, thirty amongst the refuse of one of these, attracted no doubt by the garbage which is placed in them as bait. These were much larger and finer than any I have seen elsewhere. Contrary to the comparative sizes of the two sexes, as figured by Dr. Leach, I found the males larger than the females, exceeding them in length by about half an inch. Thus, Leach's figure of the male is not at all equal in size or apparent strength to those which I found at Bognor, but that of the female is about the ordinary size of that sex. Like all the slow moving Crustacea, they are very liable to be covered with small fuci, so that they are sometimes completely concealed by a mass of these marine plants growing upon their surface, where their roots find a secure hold amongst the villous coat of the shell and limbs.* This is especially the case with the

* Say supposes that the fuci, which are found covering certain Crustacea, are merely entangled mechanically in the hooked hairs by which they are covered;

females, which in this, as in many other species, are less active than the males. Their movements are extremely slow and measured, and they are very timid, concealing themselves under the fuci, and remaining for a time almost motionless. But notwithstanding their timid and lazy character, they seize the object of their anger by a sudden and unexpected snap, and nip with great force, holding on with extraordinary firmness and tenacity, although unable, from the bluntness of their pincers, to inflict a wound. The manner of their seizing any object, when from their slow motion it is least expected, reminded me of the mode in which I have seen the *Otolicnus tardigradus* seize a bird, or other small living animal; and any one who has seen both, must, I think, be struck with the similarity.

This species of *Pisa* formed the type of the genus *Blastus* of Leach, who, however, afterwards reunited the two forms, which certainly are not sufficiently distinct to warrant their separation. It would appear from the paucity of observations which I have found of the occurrence of this species, that it is not a common one; or at least that it is very local. Mr. Couch says in his " Cornish Fauna" that it is not common in that county. Dr. Leach gives, as its localities, " The Isle of Wight, Teignmouth, and Brighton." It is not mentioned by Mr. Hailstone in his MS. Catalogue of Hastings Crustacea, which he obligingly sent me, nor do I remember to have found it there. I have taken many small specimens on the Dorsetshire Coast by dredging, and, as I before observed, in very large numbers at Bognor. The only account of its occurrence as an Irish species is, that "two examples exist in Mr. Ball's

but there is no doubt that they actually grow upon them, and are attached by roots. This is evident from the healthy state of the little plants, as well as from the direction of their branches.

collection which were obtained at Roundstone, Connemara."

It inhabits, also, the Mediterranean; and I have observed a remarkable peculiarity in some of the specimens from that locality. The anterior pair of legs, as I have before mentioned, do not assume their full size and development until the animal is quite adult; but I have seen Mediterranean specimens of a very small size comparatively, with the full adult development of the feet. In such cases we might expect to find the reproductive organs fully perfected, from some local circumstances favourable to their development, whilst the general growth of the animal had been retarded, probably by deficiency of nourishment.

DECAPODA.　　　　　　　　　　　　　　　　　　　MAIADÆ.
BRACHYURA.

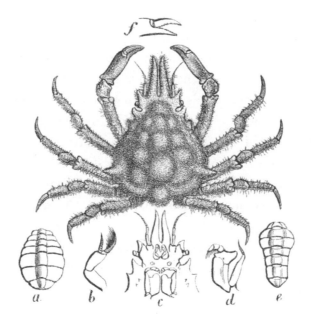

GIBBS'S SPIDER-CRAB.

Pisa Gibbsii. Leach.

Specific character.—No spines on the lateral margin. A strong spine on each branchial region, and a large prominent tubercle just above the posterior margin of the carapace.

Cancer biaculeatus,	MONTAGU, Linn. Trans. XI. t. i. f. 2. p. 2.
Pisa "	LEACH, Edinb. Encycl. VII. p. 431.
" Gibbsii,	Id. Trans. Linn. Soc. XI. p. 327. Malac. Brit. t. xix. EDW. Hist. Crust. I. p. 307. COUCH, Cornish Fauna, p. 65.

THE general form of the carapace in *Pisa Gibbsii* is very different from that of *P. tetraodon*. The rostrum is much longer, being not less than half the length of the rest of the shell, and its two horns, in the male, are parallel throughout almost their whole length; but in the female they are shorter, and divergent for about one-third of their length, as in the former species. The lateral margin of the carapace is without spines,—excepting, in some specimens, a very small one on the hepatic region. The supra-orbitar spine is smaller than in the other species, not exceeding one-third the length of the rostrum in the male; it is directed outwards and forwards; the post-orbitar spine is very small. The regions of the carapace are very strongly marked and gibbous, particularly the genital and intestinal, and they are separated by deep furrows. There is on each branchial region a strong prominent spine which, with a large round tubercle just above the middle of the posterior margin, on the intestinal region, form an obtuse triangle. The antennæ, the pedipalps, and the abdomen, are very similar to those of *P. tetraodon*. The anterior pair of feet are of moderate size, not nearly so broad and massive as those of the other species, and the hands compressed. The remaining feet are tuberculated, excepting the penultimate joint of the second pair, which is without tubercles or spines. The whole surface is covered with a very dense villous coat, much thicker than in *P. tetraodon*, and there are a few tufts of longer club-shaped hairs interspersed, with which also the base of the rostrum and that of the antennæ are furnished.

This species is exceedingly liable to the growth of foreign substances upon the surface, to which the dense villous covering affords a very ready and firm attachment. I have a specimen in my collection the form of which is almost completely concealed by a mass of sponge which has grown on its back.

Dr. Leach states that it was first noticed by Mr. Gibbs, who was employed as a collector by Montagu. It was described and figured by the latter indefatigable naturalist, in the eleventh volume of the "Transactions of the Linnean Society," under the name of *Cancer biaculeatus*; and Dr. Leach afterwards assigned to it its present name after the discoverer.

According to the same authority it is not an uncommon species on the southern coast of Devon and Cornwall. In the latter county Mr. Couch says it is not uncommon, occurring at various depths, from two to twenty fathoms. I have obtained it at Hastings; where Mr. Hailstone also mentions its frequent occurrence; and Dr. Milne Edwards mentions it as an inhabitant of the French coast.

It is generally found in deep water, and is taken either by the trawl net, or by dredging. It spawns in December, according to the observation of Mr. Hailstone.

GENUS HYAS, Leach.

Cancer,	Herbst.
Maia,	Bosc.
Inachus,	Fabricius.
Pisa,	Latr.
Hyas,	Leach, Edwards.

Generic character.—*External antennæ* with the basal portion slightly narrowed forwards, and separated from the outer portion of the orbit by a notch; the second joint dilated externally, longer than the third. *External pedipalps* with the third joint notched at the internal apex. The first pair of legs thicker than the rest, shorter than the second pair, and equal; the fingers tapering to the point, and when closed, meeting throughout nearly their whole length. The remaining pairs of legs simple, slender, long, almost cylindrical; the terminal joint without spines beneath. *Carapace* tuberculous, elongate-subtriangular, much rounder at the posterior margin; rostrum of moderate length, triangular, depressed; the laciniæ somewhat converging. The lateral margin with a strong spear-shaped process immediately behind the orbit. *Eyes* capable of being deflexed within the orbits. *Abdomen* seven-jointed in both sexes; the terminal joint in the male is transversely oval, and the corresponding margin of the penultimate joint is broadly emarginate to receive it.

This genus bears considerable relation to *Pisa*, from which it differs, amongst other characters, in the dilated form of the second joint of the antennæ, and the absence of spines beneath the last joint of the legs.

DECAPODA.
BRACHYURA.

MAIADÆ.

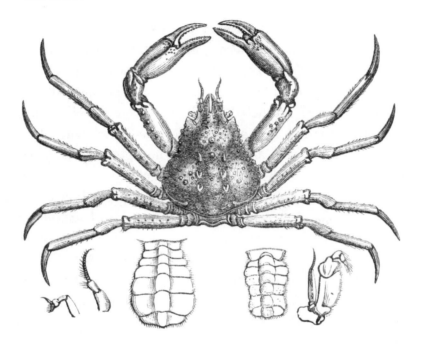

Hyas araneus.

Specific character.—Carapace not contracted behind the post-orbitar hastiform process.

Cancer araneus,   Linn. Syst. Nat. I. 1044.
,, *Bufo,*   Herbst, I. t. xvii. f. 59. p. 342.
Inachus araneus,   Fabr. Ent. Syst. Suppl. 356.
Maia ,,   Leach, Edinb. Encycl. VII. 394.
Hyas ,,   Id. l. c. p. 431. Id. Malac. Brit. t. xxi. A.   Edw. Hist. Crust. I. p. 312. Couch, Cornish Fauna, p. 66.

This is the largest British species of the family with the exception of *Maia Squinado*. The carapace is of an elongate-triangular form, the posterior margin very much rounded, and the anterior portion considerably narrowed. The rostrum is triangular, its two laciniæ nearly parallel at their inner edge, converging at the points, somewhat flattened above, and slightly hollowed beneath. The external antennæ are remarkable in this, as in the other species of the genus, for the dilated form of the external margin of the second joint, which is also considerably longer than the succeeding one; the peduncle is nearly as long as the rostrum. The eyes are but little larger than the footstalk, and capable of being retracted within the orbit, which is large and open, arched above, and protected posteriorly by a strong hastate process. There are no spines on any part of the body or limbs; but the carapace is covered with low tubercles of various sizes. Of the external pedipalps the second joint is quadrate, slightly produced at the anterior and inner angle; the third joint of an irregular form, and somewhat notched at the inner apex for the articulation of the terminal portion. The abdomen of the male is of a very peculiar form. The third joint is the broadest, the fifth and sixth nearly equal, and the latter excavated in its distal margin to receive the seventh joint, which is transversely oval, or rather reniform, being broadly emarginate at the terminal margin. The abdomen of the female is broadly oval, and has a broad tuberculated carina, which is also the case with that of the male. The body and limbs are partially covered with a villous coat.

The dimensions of a fine male are as follows:

	In.	Lines.
Length of the carapace	3	6
Breadth of do.	2	6
Length of the anterior legs	5	3

"This species," says Dr. Leach, "is very common on the coasts of Scotland and Kent. On the shores of Devonshire it is of rare occurrence." I have received it from

Worthing in Sussex, and from the coast of North Wales, through the kindness, respectively, of my friends Mr. Dixon and Mr. Eyton. I have obtained it at Hastings, where it occurs in considerable abundance; and dredged it on oyster-beds at Sandgate, of large size, at from ten to twelve fathoms.

The following particulars respecting the occurrence of this species on different parts of the coast of Ireland, are very interesting, and are taken from the Catalogue of Irish Crustacea, by my friend Mr. W. Thompson.

"Mr. Templeton has noticed this species as taken at Carrickfergus; and native specimens are in Mr. J. V. Thompson's collection. It has been obtained at Youghall and Dublin by Mr. R. Ball. We take it by dredging in the loughs of Strangford and Belfast, where, too, it is commonly thrown ashore. In the estuary, at little more than half a mile from Belfast, a number of large specimens of this Crab were captured in the month of October 1839, on the hooks attached to hand lines, much to the surprise of the fishermen, who had never met with them so near the town before, or in brackish water. The lug-worm (*Lumbricus marinus*) was the bait attacked in this instance by the Crabs. *Hyas araneus* was taken in the dredge at Bundoran, on the western coast, by our party in July 1840, and very small living specimens were found under stones, between tide-marks at Lahinch, on the coast of Clare. In Mr. Hyndman's cabinet are two Crabs of this species, with oysters attached to their backs. The oyster (*Ostrea edulis*) on the larger Crab is three inches in length, and five or six years old, and is covered with many large *Balani*. The 'shell,' or carapace of the Crab is but two inches and a quarter in length, and hence it must, Atlas-like, have borne a world of weight upon its shoulders.

The presence of this oyster affords interesting evidence that the *Hyas* lived several years after attaining its full growth. Both crabs and oysters, though dead, were brought to Mr. Hyndman in a fresh state. The hairs on the body and legs of specimens in my collection are longer in the small than in the large individual. On the north-east coast of Ireland, the *H. araneus* is very much preyed on by the codfish.

" In January 1840, I saw specimens of this Crab of very large size on the coast near Edinburgh; the carapace of one which I measured was three inches in length, and the extent from the extremities of the first pair of legs eleven inches."

Mr. Hailstone states that this Crab spawns in February: this, however, cannot be universally the case, as I took several females at Sandgate early in May, in the year 1843, every one of which was carrying her load of spawn, which is of a rich deep orange colour.

DECAPODA. MAIADÆ.
BRACHYURA.

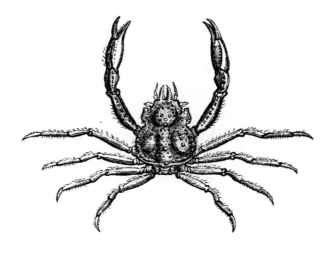

Hyas coarctatus. Leach.

Specific character.—Carapace distinctly contracted immediately behind the post-orbitar process.

Hyas coarctatus,   Leach, Trans. Linn. Soc. XI. p. 329. Id. Malac. Brit. t. xxi. B. Edw. Hist. Crust. I. p. 312. Couch, Cornish Fauna, p. 66.

This is a small and elegant species, differing so much in the contour of the shell from *Hyas araneus*, as to be distinguished at a glance from that species, although agreeing with it in almost all the essential characters. The carapace is very broad anteriorly, and suddenly contracted at the sides, both of which characters arise from the extraordinary breadth of the post-orbitar processes, which are half lyre-shaped and lamelliform. The rostrum is bifid, triangular, and each lacinia has a series of minute tubercles along the middle. The whole carapace is tuberculated. The antennæ, the eyes, the orbits, and the pedipalps, are very similar to those of the former species. The first pair of legs in the male are fully half as long again as the body; those of the female, which are slender, about the length of the body; the arms, wrists, and hands are tuberculo-carinated. The remaining legs are slender, shorter and smaller in proportion, than in *H. araneus*; the third joint with a line of small tubercles above. The abdomen resembles in each sex that of the former species. The colour of the carapace and legs above is reddish-white, the tubercles a beautiful pink or rose-colour; the under parts dirty white. The dimensions of a fine male taken at Sandgate by myself, are as follow:

	In.	Lines.
Length of Carapace	1	3
Breadth of do.		9
Length of first pair of legs	1	9

It is remarkable that in Dr. Leach's plate of this species the figure of the male is very much smaller than that of the female. In the specimens which I have taken, the contrary has been generally the rule, and the males have been much larger than those figured by him.

This species was discovered by Dr. Leach in the Frith of Forth, and afterwards found by him on the southern coast of Devon. I procured it at Hastings. Mr. Eyton sent it to me from the coast of North Wales; Mr. Couch from Cornwall, and Mr. Dixon from Worthing. Dr. Leach mentions Sandgate as a particular habitat, where I also obtained several specimens by dredging, in May. I have received it through the kindness of Mrs. Tate from Zetland, and from Orkney by Dr. Pollexfen and Dr. Duguid. As an Irish species, it has occurred at Youghall, in Dalkey Sound near Dublin; in the loughs of Strangford and Belfast, and at the Giant's Causeway. "Thus," says

Mr. Thompson, "from the North to the South of Ireland this species prevails." This extensive range authorises me to consider it as even more generally distributed on our coasts than *H. araneus*. In the young state it is very difficult to distinguish the two species, as the former has, in its early age, the spreading form of the post-orbitar processes which distinguishes the present species in its perfect adult condition, and which is gradually lost by the other. It is said by Mr. Hailstone to spawn in January. Amongst those which I obtained at Sandgate in the month of May, were several females, all without spawn.

Mr. Hailstone described in the eighth volume of Loudon's Magazine of Natural History, what he considered to be a distinct species, under the name of *Hyas serratus*. There can be no doubt that these were very young specimens of the present species, as was suggested by Mr. Westwood in some observations on Mr. Hailstone's communication. There were three specimens, which Mr. H. states were all males; but as the largest was only a quarter of an inch long, it would be impossible at so early a period to distinguish the male from the female by the abdomen.

DECAPODA.　　　　　　　　　　　　　　　　　　MAIADÆ.
　BRACHYURA.

GENUS MAIA, Lam.

    Cancer,　　Herbst.
    Inachus,　　Fabr.
    Maia,　　　Lam. Leach, Edwards.

Generic character.—*External antennæ* with the basal portion very broad, forming a considerable part of the inferior boundary of the orbit, furnished with two strong spines, the outer one directed outwards and forwards, the inner curved downwards; the moveable portion inserted at the outer and upper angle of the basal portion, where it fills the inner canthus of the orbit. *Internal antennæ* placed in triangular fossæ, between the anterior extremity of which is a strong spine, exactly similar to the inferior spine of the basal joint of the external antennæ, and ranging with them. *Eyes* not thicker than their peduncles, which are elongated and slightly curved. *Orbits* deep, oval; their upper boundary, which is arched, having two fissures. *Carapace* ovate-subtriangular, convex, covered with numerous spines or tubercles. *Rostrum* very strong, bifurcate, the horns somewhat divaricate. *Anterior legs* elongated, thicker than the others in the adult male, but much smaller in younger age, and in the female; the hands and wrists long, the fingers tapering and pointed, and scarcely toothed. *Legs* of the remaining pairs elongate, cylindrical, the terminal joint naked at the extremity, and without spines beneath. *Abdomen* seven-jointed in both sexes.

DECAPODA.   *MAIADÆ.*
BRACHYURA.

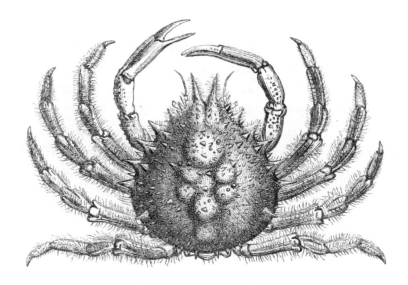

SPINOUS SPIDER-CRAB.

CORWICH.

Maia squinado.

Specific character.—Carapace convex, covered with sharp spines.

Cancer Squinado,	HERBST, I. t. xiv. f. 84–85, (jun.) Id. III. t. lvi. (adult.)
,, Maia,	SOWERB. Brit. Misc. t. xxxix.
Maia Squinado,	LATR. Hist. Nat. des Crust. VI. p. 93. BOSC, Hist. Nat. des Crust. I. p. 257. LEACH, Trans. Linn. Soc. XI. p. 326. Id. Malac. Brit. t. xviii. EDW. Hist. Crust. I. p. 327.

THE carapace of this species of Maia is considerably convex, of an ovoid form, but becoming more triangular in adult age, by the increased narrowing of its anterior portion. The rostrum is strong and prominent, its two horns somewhat diverging, so as to leave a triangular space between them. The orbit has a strong spine above its outer angle, and a smaller one at the base of the former; its superior boundary is arched and rounded. The lateral margin has five or six very strong sharp spines, the anterior of which bounds the outer angle of the orbit. The upper surface of the carapace is covered with innumerable spines and tubercles. The under surface of the anterior portion is furnished with five strong spines, two on each side on the basal joint of the external antennæ, the outer one directed forwards and outwards, the other curved downwards, and a single one at the root of the rostrum, likewise curved downwards. The second and third joints of the antennæ of nearly equal length, and inserted at the outer angle of the basal joint. Anterior pair of legs in the adult male nearly twice as long as the carapace, much larger than the succeeding ones; the arm and wrist tuberculated; the hand scabrous; the fingers very taper, pointed, the moveable one slightly curved, scarcely denticulated. The remaining legs cylindrical, without spines or tubercles; the second pair nearly half as long again as the carapace, the rest diminishing regularly to the fifth; the last joint very slightly curved, its extremity naked, abruptly smaller, and pointed. The abdomen is in each sex seven-jointed. In the male, the second joint is very narrow at the insertion of the last pair of legs, the anterior part of it becoming abruptly much wider; the sides of the remainder are nearly parallel, becoming, however, a little narrower, and the terminal margin is rounded. It has a broad carina occupying one-third of its breadth. In the female it is oval.

There are few species of Crustacea in the form of which age produces so great a change as in this. The younger individuals not only exhibit the more slender and shorter dimensions of the anterior legs, but the anterior part of the carapace is much broader in proportion; a character which permanently belongs to the Mediterranean species, *M. verrucosa*.

Pennant's figure of what he terms *Cancer maia* belongs to *Lithodes arctica*, and it is very probable that he, as well as others, has confounded these two species, before the true characters of Crustacea were understood, and indeed before naturalists in general were aware of the value of specific characters.

There is a species found in the Mediterranean very nearly allied to this, and which has been supposed to inhabit our southern coast. It is the *Maia verrucosa* of Edwards already alluded to: it is readily distinguished from this by the absence of spines on the surface, which are replaced by tubercles; by the greater extent and development of the supra-orbitar arch; by the breadth of the anterior portion of the carapace, which remains to the adult age as broad as in the younger state; and by the depressed form of the carapace. I believe *M. verrucosa* has not been taken on our shores; those found in Cornwall, and considered as such by Mr. Couch, being undoubtedly the present species.

This Crab is found in great abundance on almost all parts of our southern and western coast. In Ireland it occurs also on the southern coast. It is by far the largest species of the family, and with the exception of the great Crab, *Cancer pagurus*, the largest of the British Brachyura. I have a specimen taken in Plymouth Sound, the carapace of which is eight inches in length, and nearly six in breadth, and the length of the anterior feet is fifteen inches.

It is eaten by the poorer classes, though I understand it is but indifferent food. Like all the other triangular Crustacea, the fishermen inveterately term it "spider;" and they appear to have very little idea of any affinity between these forms, and the Crabs properly so called. I remember some years since seeing in one of the back streets of Poole, near the water-side, a little girl standing by a small table, on which was a plate containing two of these Crabs, of moderate size, cooked and for sale. On my accosting her with " Pray do they eat these crabs here ?" She replied with a look of great surprise at my ignorance, " They ben't crabs, sir, them's spiders !"

Mr. Richard Couch informs me that in Cornwall several dozens of "the Corwich" are sold for sixpence, but that they are more frequently given away to those who ask for them. Mr. Couch adds, that he never saw a soft one, or one soon after casting its shell, although they are often taken "*peel*," or ready to cast it. This, doubtless, arises from the extreme secrecy of their retreats when undergoing this process.

The following account, for which I am indebted to the gentleman just mentioned, is very interesting, and it affords another opportunity of confirming the true metamorphosis of the decapodous Crustacea. "This is the most abundant of all the Crabs found on our coast, but it does not make its appearance so early in the season as the Common Crab, the Lobster, or indeed any other ; it is rarely found earlier than May, but from that time till the end of the fishery in August or September, these Crabs make their appearance in vast numbers, to the great vexation of the fishermen; for it is found that from the time these begin to enter the pots, the more valuable kinds considerably decrease in number ; and this is supposed to arise from their restless activity. No sooner are they in the crab-pot, than they are continually in motion, scrambling from one part to another, and in this way frighten the Crab and Lobster, and prevent them from entering. In the spring and early part of the summer they lie concealed beneath the sand, in deep water. About May they leave their places of concealment, but never come into shallow water, as does the Common Crab; the latter is often found in crevices of rock, or beneath stones left by the receding tide; but this is never the case with the *Corwich*. They shed their spawn about August or September, at some short distance from the shore, most probably in the sands. In this, too, they differ from the Common Crab, for even when the spawn is quite mature for "casting," they enter the pots as readily as at any other time; whilst on the other hand it is a very rare occurrence to catch the Common Crab with spawn, unless it be with a dredge-net. It would seem either that they grow very fast, or that the young differ considerably in their habits from the larger ones; for whilst it is very common to find specimens measuring nine or ten inches in the length of the carapace, it is very rare indeed to get one less than three inches; and a fisherman tells me that after many years' fishing he caught one about the size of a half-crown, which was the smallest he ever saw.

"The ova, when quite ready for shedding, (fig. 1,) are about the size of a very small mustard-seed, and of a reddish-brown colour, besprinkled with small dark spots. After keeping them suspended in sea-water for twenty-four hours, some of the ova dropped from their attachments, and soon after the young escaped, and this evidently by their own exertions, as distinct motions were easily observable under the microscope while they were yet enclosed. When they first escape, they are, as it were, rolled on themselves, (fig. 2,) the caudal extremity being bent on the body; but this is soon changed for the position represented in fig. 3. I could detect no spine on the anterior part of the carapace, which was quite smooth, but marked with dots. The eyes are sessile and large; the claws, particularly towards the extremity, covered with minute hairs."

These interesting observations of Mr. Richard Couch afford a fresh confirmation of the truth of the metamorphosis of the brachyurous Crustacea, and it is to be hoped that whenever an opportunity occurs to any observer to preserve and examine the embryo, and the subsequent progress towards the perfect state of any other species, similar notes may be made, and thus we shall hereafter arrive at a knowledge of this curious process in most of our native species.

An ordinary sized Corwich, as Mr. Couch informs me, bears at one time upwards of seventy-six thousand eggs.

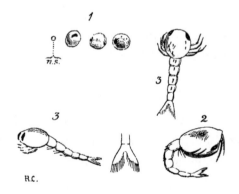

GENUS EURYNOME, Leach.

Cancer,	Pennant.
Eurynome,	Leach, Risso, Edwards.

Generic character.—*External antennæ* scarcely longer than the rostrum; the basal joint triangular, and perfectly united to the surrounding parts; the second inserted at its apex, at the inner canthus of the orbit, and beneath the rostrum; second joint larger, but shorter than the third. *External pedipalps* with the third joint dilated at the outer and emarginate at the inner angle. *Anterior pair of legs* in the male, larger and much longer than the succeeding ones; hands long, linear; fingers inflected. The second to the fifth pairs of legs linear, diminishing regularly in length. *Carapace* irregularly rhomboidal, produced anteriorly, and much rounded behind, verrucose. *Rostrum* bifid, the laciniæ triangular, flattened, slightly divaricate. *Orbits* deep, above strongly arched, with a single fissure near the external angle. *Eyes* retractile, globular, larger than the peduncles, which are short. *Abdomen* seven-jointed in both sexes.

This genus is the only British representative of a highly interesting and curious, as well as natural family, agreeing nearly with the genus *Parthenope* of Fabricius, and comprising a number of bizarre forms, which have for the most part very long arms and rough, rocky-looking bodies. They form upon the whole, as Milne Edwards has observed, a passage from the triangular families, to the more typical Canceridæ; and, like many other small osculant or intermediate groups, exhibit many diverse and somewhat isolated forms. Of these the present genus, *Eurynome*, may be considered as the most nearly related to the Maiadæ, with which family it agrees in the union of the basal joint of the external antennæ with the parts surrounding it, as well as in the general form of the body.

DECAPODA.
BRACHYURA.

Eurynome aspera. Leach.

Specific character.—Rostrum less than one-fourth the total length of the body. Carapace covered with numerous small warty tubercles, regularly disposed.

Cancer asper,	PENNANT, Brit. Zool. IV. t. x. f. 3. p. 13.
Eurynome aspera,	LEACH, Malac. Brit. t. xvii. EDW. Hist. Crust. I. p. 351.
„ *spinosa,*	HAILSTONE in Mag. Nat. Hist. VIII. p. 549.

THE carapace of this very pretty Crab is irregularly rhomboidal, the anterior triangle being longer than the posterior, which latter is somewhat rounded; the rostrum is less than one-fourth the whole length of the carapace, bifurcate, the laciniæ somewhat divergent, acute, and flattened. There is a large triangular laminar tooth at the outer angle of the orbit, and there are three smaller ones at the lateral margin of the branchial region. The carapace is covered with numerous small, round, warty tubercles, which, on a close examination, are found to be distributed with perfect regularity. The most conspicuous of these are two on each branchial region, and one on the centre of the cardiac. The latter, which is smooth and polished, is surrounded by ten others, which are warty, arranged in an oval form, five on each side. The external antennæ are not longer than the rostrum. The basal joint, as in the *Maiadæ*, is soldered to the surrounding parts; in which respect it differs from that of some other genera of the family in which it is detached; it is triangular, and the moveable portion is inserted at its apex, and does not extend beyond the rostrum. The second and third joints are oval, and nearly equal. The external pedipalps have the second joint oblong-quadrate; the third has the outer angle produced, and the inner angle truncate and emarginate. The anterior legs in the male are nearly twice as long as the body, and much larger than the succeeding ones, the arms and hands long, the wrists short, the fingers long and inflected. In the female they are but little larger, and scarcely longer than the second pair. The whole are covered with tubercles. The abdomen in the male is tuberculated and carinated; the terminal joint triangular. In the female it is oval, carinated, and the margin broadly ciliated.

The length of a very fine male specimen is about nine lines, and its breadth seven lines. Its colour is a light rose, intermixed with a slight tint of blueish-grey.

The *Eurynome aspera*, which is one of the rarer of the British Crustacea, inhabits deep water, having been dredged in seventy fathoms. It has been taken by dredging, or by the trawl on the coasts of Cornwall, Devonshire, Dorsetshire, and Sussex. I find by my own notes that I took a specimen in Swanage Bay, in Dorsetshire, some years since, but it has been lost. It has also been dredged off the Isle of Man, and in Loch Fyne, by Mr. McAndrew, to whom I am indebted for specimens from both localities. I have been favoured with another specimen, also a female, and loaded with spawn, by Mrs. Griffiths, who took it at Torquay. I cannot doubt that *Eu. spinosa* of Mr. Hailstone, described in the eighth volume of the Magazine of Natural History, is the young of the present species; it was taken at Hastings " in a mass of *Filipora filigrana*."

The following account of its occurrence as an Irish species, is taken from Mr. W. Thompson's Catalogue of the Crustacea of Ireland. " Marked as Irish in Mr. J. V. Thompson's collection. It is rather a rare species, and an inhabitant of deep water." In Strangford Lough several specimens were taken by Mr. Thompson and Mr. Hyndman. It has occurred in Belfast Bay, on the Dublin coast, and at Roundstone on the western coast. It was obtained also by Captain Beechey off the Mull of Galloway, at seventy fathoms. It is found on the coast of France, from whence I have received specimens through the kindness of my friend Dr. Milne Edwards.

Being found only in deep water, but little is known of its habits. The eggs are of a beautiful orange colour; they are deposited in June, or the early part of July, as I have a female specimen taken at the latter end of June, in which the eggs are so fully developed, that the embryo can be seen through the investing membranes.

When Dr. Leach established this genus, the present was the only species known. Risso has, however, since that, described another species, to which he gave the name *Eu. scutellata*,* but so imperfect is the description, that Dr. Milne Edwards found it impossible to judge, with any degree of certainty, whether it belonged to this genus or not; and if so, whether it might not be identical with the present. I

---

* Risso, Hist. Nat. de l'Eur. Merid., IV. p. 21.

possess, however, a pair of this beautiful little species from the Bay of Naples, and find it to be very distinct from ours in several points, yet bearing a near affinity to it. As a sufficient distinctive character of it has not yet been given, for that of Risso is altogether useless, I thought it desirable to notice it on the present occasion. It differs from the English species by its longer rostrum, which equals one-third of the total length, and by the absence of the scattered tubercles by which that is distinguished, instead of which there are several broad, flat, shield-like elevations. Risso had not seen the female, of which sex I have a specimen loaded with eggs of a deep amber colour.

My friend Professor Forbes dredged *Eu. aspera* at a depth of thirty fathoms off the Isle of Man, and at seventy fathoms in the Ægean. This evidently shows that the genus belongs to deep water, an observation which holds good of all the family of the *Parthenopidæ*.

GENUS XANTHO, Leach.

Cancer,	Montagu, Herbst.
Xantho,	Leach, Edwards.

Generic Character.—*External antennæ* very short, the basal joint longer than it is broad, in contact with the front only at its anterior internal angle; the moveable portion inserted at the inner canthus of the orbit; the second joint considerably larger than the succeeding ones. *Internal antennæ* placed obliquely immediately under the front. *External pedipalps* with the third joint quadrate, the inner anterior angle truncate and slightly emarginate. *Carapace* very broad, slightly convex from before backwards; the latero-anterior margins with the front forming a semi-ellipsis; the latero-posterior margin nearly straight; front projecting, divided by a slight fissure; *orbits*, with a fissure beneath, at the external angle. *Anterior legs* very large, nearly equal, the fingers pointed. The *posterior pairs* short, compressed; the terminal joint very short. *Abdomen*, in the male, five-jointed; in the female, seven-jointed.

DECAPODA.
BRACHYURA.

CANCERIDÆ.

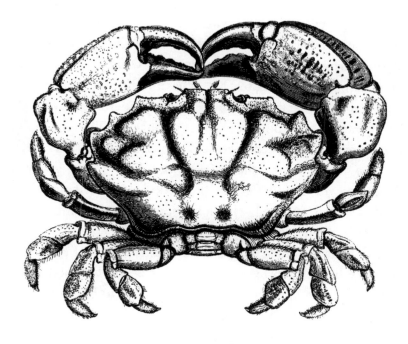

Xantho florida. Leach.

Specific character.—Carapace deflexed anteriorly; latero-anterior margin with four strong obtusely triangular teeth; fingers black, without grooves; the second to the fifth pairs of legs with the third joint only ciliated on the upper edge.

Cancer floridus,	Montagu, Trans. Linn. Soc. IX. t. ii. f. 1. p. 85.
,, *incisus,*	Leach, Edinb. Encycl. VII. p. 391.
Xantho incisa,	Id. l. c. p. 430.
,, *florida,*	Id. Trans. Linn. Soc. XI. p. 320. Malac. Brit. t. xi.
,, *floridus,*	Edwards, Hist. Crust. I. p. 394.

The carapace of *Xantho florida* is about two-thirds as long as it is broad; the anterior portion somewhat deflexed; the latero-anterior margin with four strong obtusely triangular teeth, and reaching nearly as far back as the anterior part of the cardiac region. The surface of the anterior portion has several broad flattened elevations, which are separated by grooves, the principal of which are continuous with the intervals between the lateral teeth; the posterior portion nearly smooth. The front is very slightly waved, and sub-emarginate. Orbit with a fissure at the inner angle beneath. The anterior legs very large and strong; the wrist with a double tubercle above; the hand rugous, the fingers without grooves. The remaining legs short, slightly compressed, the third joint only hairy on the upper edge, the fourth and fifth joints grooved. Abdomen in the male five jointed, in the female seven jointed; oval, ciliated with long hairs. The colour of this species is a reddish brown, the claws black.

The male is much larger than the female, and his claws are very large in proportion to the size of the body. A full-sized male is more than an inch and a half long, and nearly two inches and a half broad; and the anterior legs of such an individual are nearly four inches long, and the hand is three-quarters of an inch broad.

This species formed the type of a new genus established by Dr. Leach, and was at that time the only one known to him. Since that time, however, many others, some before known and placed in other genera, and some since discovered, have been ascertained to belong to it, so that it now consists of between twenty and thirty species, inhabiting every quarter of the world. Until lately, however, it has been considered our only indigenous species. It was first described by Montagu in the "Transactions of the Linnæan Society," under the name of *Cancer floridus;* but, as Dr. Leach very truly says, he must have been misled in supposing it to be identical with Linnæus's species of the same name. The *Cancer floridus* of Herbst, which Montagu quotes also as a synonyme of this species, is a characteristic figure of *Zozymus æneus*.

It is found in considerable numbers on the coast of Cornwall and Devonshire, and also in Dorsetshire. It has been observed on several parts of the coast of Ireland. The female produces a large quantity of eggs, which are of a reddish brown colour. Of its peculiar habits nothing is known.

DECAPODA.
BRACHYURA.

Xantho rivulosa.

Specific character.—Carapace nearly horizontal; latero-anterior margin with four triangular teeth; fingers brown, the moveable one grooved above; the second to the fifth pairs of legs with all the joints ciliated on the upper edge.

Cancer hydrophilus,   HERBST, I. t. xxi. f. 124. p. 266.
Xantho florida, var. β,   LEACH, Trans. Linn. Soc. XI. p. 320.
„   *rivulosus,*   EDWARDS, Hist. Crust. I. p. 394. Roux, Crust. Mediterr. t. xxxv. COUCH, Cornish Fauna.

THIS species exceedingly resembles *X. florida,* and has been doubtless often confounded with it. There are, however, numerous well marked distinctive characters, as the following description will show on a comparison with that of the former.

The carapace is nearly horizontal, the anterior portion

being very slightly deflexed; the front nearly straight, projecting, the margin minutely beaded. The latero-anterior margin with four triangular teeth, the posterior of which reaches scarcely beyond the line of the posterior edge of the gastric region; the inequalities of the surface and the intervening grooves, are not very strongly marked. The anterior legs are large and strong, the wrist bituberculated; the moveable finger has a distinct groove on the outer side of the upper surface, extending from the joint nearly to the extremity. The remaining pairs of legs are compressed, and the upper edge of all the joints ciliated.

The general colour is yellowish, with red markings; the fingers brown, sometimes but little darker than the rest of the shell. The specimens which I have seen have all been smaller than the full size of *X. florida*.

It appears, then, that the present species differs from the former in the following particulars:—the carapace is much more horizontal, the inequalities of the surface less conspicuous, the lateral teeth more angular, the front less deflexed, its margin short and prettily beaded, which is not the case with those specimens of *X. florida* which I have examined. But besides these comparative characters by which the two species may be distinguished, when examined together, there are others of a positive kind by which the present animal may be readily detected. The moveable finger is grooved; the whole of the joints of the legs are ciliated on the upper edge, whereas in *X. florida* this is the case only with the third joint. The colour of the pincers in this species is brown, in the other it is quite black.

There can be no doubt that this is the *Cancer hydrophilus* of Herbst. It is figured also by Savigny in the "Crustacea of Egypt;" it occurs in Risso's "Crustacea of the Neighbourhood of Nice;" and Edwards says that it inhabits the western coast of France. In all probability it is identical with *X. florida,* " var. β digitis concoloribus" of Leach; but Mr. Couch of Polperro was the first to detect it as an English species, and to refer it to its proper name; and it was also detected by Captain Portlock as an Irish species, a specimen having been obtained at Portruch, in the county of Antrim. I have been favoured by Mr. Couch with specimens from Cornwall; I have also received it from North Wales, through the kindness of my friend Mr. Eyton. Mr. Couch, writing from Polperro says, "*Xantho rivulosa* is common with us, rather more so than *X. florida.* It is found concealed under stones at low-water mark; is of rather slow habits, and exuviates much in the same manner as the common crab." There is indeed but little difference in this respect amongst all the true brachyurous forms.

GENUS CANCER.

CANCER,	Linn. Leach, Bell.
PLATYCARCINUS,	Latr. Edwards.

Generic Character. — *External antennæ* with the basal joint very long and thick, filling the hiatus between the inner canthus of the orbit and the front, and terminating forwards in a strong, angular, tooth-like projection, directed forwards and slightly inwards, reaching a little beyond the frontal line; the terminal portion is very short and slender, and arises from the internal part of the basal joint, nearer to the cell of the internal antennæ than to the orbit. *Internal antennæ* directed forwards, placed in longitudinal cells. *External pedipalps* with the third joint excavated at the anterior and inner margin. *Anterior feet* nearly equal, robust; the others, more or less hairy, but without spines. *Carapace* transversely elliptic, somewhat elevated, with the regions obviously marked; front trifid; *orbits* with a strong tooth over the inner canthus; and with two fissures above, and one beneath: the latero-anterior margin on each side extends back to the centre of the cardiac region, and passes off into a sinuous, granulated ridge, which rises over the latero-posterior margin; it is divided into ten lobes, of which the last is very small, and often obsolete. *Eyes* placed on short peduncles. *Abdomen*, in the male, five-jointed; in the female, seven-jointed.

This genus is readily distinguished from its immediate congeners by the form of the basal portion of the external antennæ, by the direction of the internal antennæ, and by the form of the latero-anterior margin of the carapace, which is, in this genus, uniformly ten-toothed. There is but one species of this genus, as now restricted, native of the shores of this country, or indeed of Europe, all the others being South American.

The generic name *Cancer* was applied by Dr. Leach to this, the present genus, as restricted by him; and I have elsewhere* stated my reasons for restoring it, after Latreille had, in the French Museum, assigned to it the name of *Platycarcinus*, in which he had been followed by Dr. Milne Edwards. When the characters of the present genus were first defined, the only known species was the common large eatable crab of our coasts, the *Cancer Pagurus* of Linnæus. Subsequently another species was added by Say, and since that three others by myself, from the South American collection of Mr. Cuming. The whole of these are described in a monograph of the genus just referred to.

* " Transactions of the Zoological Society," vol. i. p. 335.

DECAPODA.   *CANCERIDÆ.*
BRACHYURA.

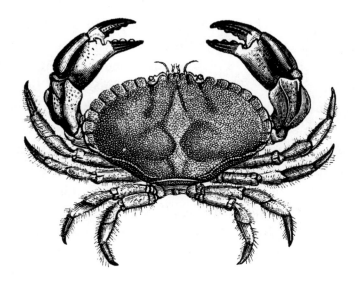

GREAT CRAB.

Cancer Pagurus. Auct.

Specific character.—Shell granulated; latero-anterior margin ten-lobed, the lobes contiguous, quadrate, entire; hands smooth.

Cancer Pagurus,	LINN. Syst. Nat. XII. i. 1044. HERBST, Krab. t. ix. f. 59. PENN. Brit. Zool. IV. t. iii. f. 7. LEACH, Malac. Brit. t. x. BELL, Trans. Zool. Soc. I. p. 341.
Platycarcinus Pagurus,	EDWARDS, Hist. Crust. i. p. 413
Jun. *Cancer inciso-crenatus,*	COUCH, Cornish Fauna, p. 70.

THE carapace is transversely oblong, flattened, slightly elevated in the middle, somewhat rounded before and behind; the surface minutely granulated, smooth, with the regions but slightly marked. The latero-anterior margin is slightly recurved, divided into ten quadrate lobes, the sides of which are contiguous, and the margins entire;

the last lobe inconspicuous, and passing into the posterior marginal line, which terminates immediately anterior to the posterior marginal ridge. The front trifid, the teeth nearly equal. The orbits are round, with a strong triangular tooth over the inner canthus, which does not project as far as the front, and a smaller one between the two superior fissures. The external antennæ have the basal joint much elongated, and terminating forwards in an obtuse tooth; the first joint of the moveable portion club-shaped, the second cylindrical. The internal antennæ stand forwards, the anterior half being folded directly backwards when at rest. The sternum minutely punctated, and furnished with small patches and lines of short scanty hair. The abdomen in the male, has the margin fringed with short hair, and the surface with numerous small tufts of short stiff hair; the last joint forming an equilateral triangle: in the female the sixth joint is very large, the terminal one triangular, the sides slightly sinuated. The anterior pair of feet large, robust, smooth, without spines or tubercles, minutely granulated; the hand rounded, without any ridge; the fingers with strong rounded teeth. The remaining feet slightly compressed, irregularly angular, and furnished with numerous bundles of stiff hairs.

The colour above reddish brown, in younger individuals with a purplish tint; the legs more red; the claws black; beneath nearly white.

There can be little doubt that this species was the one known to the Romans by the name of *carabus*, from whence our common name crab.* Pliny, in enumerating

---

* The common name of the wild apple has probably no reference to the animal; it is, doubtless, as Skinner has it, from *schrabben*, A. s., to scrape, to bite, from the harsh, rough taste of the fruit.

the different kinds of " cancer" says,—" Cancrorum genera carabi, astaci, maiæ, paguri, heracleotici, icones et alia ignobiliora." It would appear by this passage that the term *Cancer* was applied to the whole of the Malacostracous Crustacea; for not only are the *brachyura* and some of the larger *macroura* evidently here designated, but the " alia ignobiliora," in all probability, indicated all the smaller and less important forms.

The habits of this species have been perhaps more thoroughly investigated, and are better understood than those of most other species. Its large size, and the excellence of its flavour, occasion it to be more sought after as an article of food than any other of the brachyurous species; and hence its habits and the places of its resort have been necessarily much observed by those whose occupation it is to procure it for the market; whilst the naturalist has found it a convenient species for his more scientific investigations, whether as it regards its history or its structure.

It inhabits the whole of our coasts, preferring those parts which are rocky; and its usual retreats are amongst the holes in the rocks, where it generally retires when not engaged in seeking its food. It is often seen in such situations, even when the tide has retreated sufficiently to render the rocks accessible, as, for instance, among those on the shore at Hastings, where I have often seen them in the pools and caverns, left by the receding tide. These are, however, always small individuals, rarely more than three inches in breadth; the larger ones remain farther at sea amongst the rocks in deep water; and they also bury themselves in the sand, but always in the immediate neighbourhood of the rocks. The food of this species, like that of most others, consists principally of animal matter, such as dead fish, and the like; and it is exceedingly probable that the crabs discover their food rather by the smell than by sight, or, at least, by an impression made by the diffusion of odorous particles emanating from it, and diffused through the water. Thus they detect the bait which is often placed in such situations that it cannot be seen by them at any distance, and which consists generally of pieces of fish, in which decomposition has already commenced. Mr. Couch, indeed, states in his "Cornish Fauna," that " It is found that the freshest (bait) only will attract the crab, whilst for the lobster it is best when hung for several days to become tainted." And this may doubtless be true to a certain extent; but I have often seen crabs taken with lobsters in pots in which the bait was far from being sweet. The period of life at which the " Bon crab," as the female of this species is termed along the western coast, begins to breed is, according to Mr. Couch, when the carapace is about three inches across. The male seeks the female at various seasons; but it would appear that in this, as in the case of the *Carcinus mænas*, this often takes place immediately after her exuviation, and that the male watches for the completion of this process, when the female is in a soft and unprotected state. My friend Mr. Richard Couch, thus writes to me on this subject. " When the female retires for exuviation, she is generally accompanied by a male; and when the shell is removed impregnation takes place. If the male be discovered and removed, another will be found to have taken his place after the following tide, and this will be repeated for many times in succession." The spawn is carried by the parent for a considerable period, and is deposited "at all seasons of the year," according to Mr. Couch; Mr. Hailstone says in March; but it is most probable that it occurs during the spring and summer, as is the case with so many other species.

It was in the month of June, 1826, that Mr. J. V. Thompson* " had the good fortune to succeed in hatching the ova of the common crab," and thus, by perfect and satisfactory observation, demonstrated the theory which his investigation of Zoea had already suggested to his mind, of the true metamorphosis of the crustacea; a discovery which may rank amongst the most interesting and important that have been made within the sphere of the sciences of observation, not only in the present, but in any previous age. The extreme difficulty of preserving these little animals alive, and ensuring them a supply of their proper food, has prevented the observations of their subsequent growth from being so satisfactorily carried out as could have been wished; but the doctrine thus established has been confirmed in so many instances by observations on other species of crustacea, that the metamorphosis of these animals may now be considered as a fixed and incontrovertible truth.

The fishery for these crabs constitutes an important trade on many parts of the coast. The numbers which are annually taken are immense; and as the occupation of procuring them is principally carried on by persons who are past the more laborious and dangerous pursuits of general fishing, it affords a means of subsistence to many a poor man who, from age or infirmity, would be unable without it to keep himself and his family from the workhouse. They are taken in what are termed " crab-pots ;" a sort of wicker trap, made, by preference, of the twigs of the golden willow, (*Salix vitellina,*) at least, in many parts of the coast, on account, as they say, of its great

* See his " Zoological Researches," No. I. p. 9.

durability and toughness. These pots are formed on the principle of a common wire mouse-trap, but with the entrance at the top; they are baited with pieces of fish, generally of some otherwise useless kind, and these are fixed into the pots by means of a skewer. The pots are sunk by stones attached to the bottom, and the situation where they are dropped is indicated, and the means of raising them provided, by a long line fixed to the creel, or pot, having a piece of cork attached to the free end of the line: these float the line, and at the same time serve to designate the owners of the different pots; one perhaps having three corks near together, towards the extremity of the line, and two distant ones; another may have one cork fastened cross-wise; another two fastened together, and so on. It is of course for their mutual security that the fishermen abstain from any poaching on their neighbour's property; and hence we find that stealing from each other's pots is a crime almost wholly unknown amongst them. It is at Bognor, and Hastings, and in Studland and Swanage Bays in Dorsetshire, that I have principally had opportunities of personal observation on these points; and I am also indebted to my friend Mr. Richard Couch for some interesting observations on this subject; in addition to which I would refer to an excellent account of the crab and lobster fishery, in the 6th volume of the Penny Magazine.

Mr. Richard Couch informs me that on the coast of Cornwall "most of these crabs are sold to the lobster smacks; but, that when brought on shore for sale, those measuring six inches across the carapace are sold for twopence each; those of eight or ten inches, threepence, and the largest from sixpence to eightpence!" If the crabs are not immediately wanted on being taken out of the pots, they are placed in store pots, which are of the same form and materials as the others, but considerably larger. They are conveyed to great distances, as far, for instance, as from the coast of Norway to the Billingsgate Market, in well boxes, which are of wood, very strongly constructed, and with holes in all the sides to admit of continual change of water, as the boxes are drawn through the sea, attached to the vessel.

The male Crabs are esteemed the best for the table; they are generally larger than the females, and the claws are much heavier. They often weigh eight or nine pounds, and sometimes as much as twelve pounds.

Examples are not few of the occurrence of different species of Crustacea in armorial bearings. Prawns, Crayfish, Lobsters, and Crabs, are occasionally found, and these, not only as "canting" bearings, or puns upon the name of the bearers, but often as examples of that emblematical allusion in which the heralds of former times so much delighted. This is not, perhaps, the place to enter into much serious disquisition on the utility of such a custom; and yet one can scarcely read the quaint, but wholesome moralities, of good old Guillim, and other professors of the gentle science, without some misgivings that the matter-of-fact and prosaic scorn of such emblems, which has succeeded to the more poetical—may we not also say the happier credulity of olden time, may have given us no equivalent advantage for the loss of those striking and epigrammatic maxims. I shall venture, therefore, to indulge an old fondness for this ancient, and really not uninteresting "science," (I do not use the term in its modern and critical sense,) by giving some occasional examples of CRUSTACEAN HERALDRY. And in doing this I cannot but refer to Mr. Moule's "Heraldry of Fish," as a work not less interesting in its historical and technical details, than tasteful and elegant in its illustration. We will presume, and it appears extremely probable, that the Heraldic Crab is the present universally known and useful species. Mr. Moule observes, "The Crab, the emblem of inconstancy, appears on a shield of Francis I., one of the finest specimens of art in the collection of armour at Goodrich Court; and, according to Sir Samuel Merrick, the Crab was intended as an allusion to the advancing and retrograde movements of the English army at Boulogne, under the celebrated Charles Brandon, Duke of Suffolk, in 1523." A golden Crab, according to the same authority, was one of the cognizances of the Scrope family, and is found on the portrait of Henry, Lord Scrope. "The Crab also appears as a crest on the seals of several members of this noble family."*

The families of Bridger of Sussex, Crab of Scotland, Bythesea of Kent, and some others, also bear this animal in their coat-armour.

* Moule's "Heraldry of Fish," p. 231.

DECAPODA.
BRACHYURA.

CANCERIDÆ.

GENUS PILUMNUS

CANCER, Linn., Pennant, Herbst.
PILUMNUS, Leach, Edwards.

Generic Character.—*External antennæ* long and setaceous; the basal joint not continuous with the surrounding parts, but separated by a distinct line, and filling the inner canthus of the orbit; second joint nearly as broad as it is long, and moveable with the remaining portion; third joint longer than the second. *Internal antennæ* with the last point of the peduncle club-shaped. *External pedipalps* with the third joint transversely quadrate, the antero-internal angle emarginate. *Anterior pair* of *feet* unequal, robust, rounded; the remaining pairs rounded above, flattened beneath; the second pair not longer than the third or fourth. *Carapace* convex, the anterior part much curved from before backwards; the surface even; the latero-anterior margin extending backwards as far as the posterior part of the gastric region; *front* slightly prominent; *orbits* elliptical, the inferior margin spinulose. *Abdomen*, in each sex, seven-jointed; in the male, the third joint the broadest, the succeeding ones diminishing regularly to the apex; in the female, all the joints sub-equal.

DECAPODA.
BRACHYURA.

CANCERIDÆ.

Pilumnus hirtellus. Leach.

Specific character.—Superior margin of the orbit not spinous, but, with the front, minutely denticulated; latero-anterior margin armed with four spines (exclusive of the external angle of the orbit); hands slightly tuberculated.

Cancer hirtellus,	LINN. Syst. Nat. I. 1045. PENN. IV. t. vi. f. 1. p. 9.
Pilumnus „	LEACH, Trans. Linn. Soc. XI. p. 321; Malac. Brit. t. xii. EDWARDS, Hist. Crust. I. p. 417.

THE carapace is smooth, anteriorly much incurved; its length to its breadth, as seven to ten; the front broad, finely toothed, divided in the centre by a deep fissure; the latero-anterior margin evenly arched, furnished, exclusive of the outer angle of the orbit, with four strong sharp spines, the anterior two being frequently bifid; the hinder one the strongest, and in a line with the posterior part of the gastric region. The upper margin of the orbits very minutely toothed; the lower margin spinous, and in each a small fissure. The anterior pair of legs are remarkably strong, thick, and rounded; they are somewhat unequal, in some the right, in others the left being the larger; the wrist is tuberculated, and furnished with a single spine, and is slightly hairy; the smaller hand is tuberculated on its upper and outer surface, the larger one almost entirely smooth; the moveable finger much curved, the fixed one triangular, and strongly toothed. The remaining legs are slightly rounded above, flattened beneath; they are covered with numerous hairs, and there are also a few on the wrist and on the anterior part of the carapace, which is also covered with short down. The abdomen in the male is broadest at the proximal margin of the third joint, thence diminishing regularly to the extremity, the third to the seventh, thus forming a long acute triangle. The abdomen of the female is of the form of a long ellipse, with the proximal portion truncate; its margin is fringed with long hair. The colour of most individuals is brownish red, with obscure yellowish spots; the anterior legs brownish red, the fingers light brown; the remaining legs red, with obscure yellowish bands. In many the brownish red colour is replaced by a dull purple.

	In.	Lines.
Length of the carapace	0	7
Breadth of do.	1	0

The present species is the only one of the genus found on our coast, and it may be readily distinguished from all the foreign species by the absence of spines on the superior margin of the orbit. The figures in Dr. Leach's great work are very inferior, and would scarcely serve to distinguish it, were any of the other species indigenous to this country with which it might be compared. They must have been taken from immature specimens; but even of such they form but very erroneous representations.

It is a common species on all the western coast of England, having been taken in Cornwall, and along the coast of Devonshire, Dorsetshire, Hampshire, and Sussex. Dr. Leach mentions it being taken under stones at low tide, but those which I have obtained have been from deep water. I have dredged them in Swanage Bay, Dorsetshire; but the finest specimens I ever saw, I procured from prawn and lobster pots at Bognor, in September, 1842. It is worthy of remark, that amongst twenty or thirty specimens, I found only one female, a dead and mutilated one. It would appear from Mr. Thompson's Catalogue to be widely distributed on the coasts of Ireland, although occurring in small numbers.

The different species of this genus are very widely distributed. They inhabit the Mediterranean, the Red Sea, the East Indies, and other parts of the coast of Asia, Australia, and both the eastern and western coasts of South America.

GENUS PIRIMELA.

CANCER,	Montagu.
PIRIMELA,	Leach, Desmarert, Edwards.

Generic Character.—*External antennæ* nearly half the length of the carapace; the basal joint short, filling a space at the inner angle of the orbit; the moveable portion inserted at its inner canthus. *Internal antennæ* lying somewhat obliquely in their cavities, which open immediately under the margin of the front. *External pedipalps* extending forwards beyond the oral cavity, and covering the epistome; the third joint sub-quadrate, emarginate at the inner margin, about one third from the anterior angle, for the articulation of the palpes. *Anterior legs* small, compressed; the remaining pairs of moderate length, much compressed; the terminal joint nearly straight. *Carapace* nearly as long as it is broad, convex, with numerous strongly-marked elevations; the anterior margin arched, the posterior much narrowed; front tridentate, the middle tooth the longest. *Orbits*, with two fissures above. *Eyes*, not thicker than their peduncles, which are very thick at the base. *Abdomen*, in the male, five-jointed; in the female, seven-jointed.

Of this genus one species only is at present known. It differs from all the other *Canceridæ*, in the circumstance that the external pedipalps, instead of being confined to the opening of the oral cavity, are advanced over the epistome to the antennary cavities.

In its affinities this genus probably approaches the *Portunidæ* by the genus *Carcinus*; possibly *Panopæus* may be intermediate between them.

DECAPODA.
BRACHYURA.

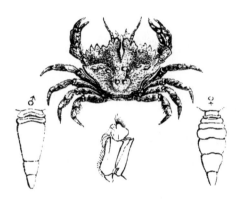

Pirimela denticulata.

Cancer denticulatus,   Montagu, Trans. Linn. Soc. IX. p. 87. t. ii. f. 2.
Pirimela denticulata,   Leach, Malac. Brit. t. iii.   Edwards, Hist. Crust.
                        I. p. 424.

THE general form of this pretty Crab will at once strike us as differing very greatly from all those which have preceded it. The carapace is very little broader than it is long; the anterior margin is so much arched, as to form nearly a semicircle, whilst the posterior portion is regularly and greatly narrowed. The latero-anterior margin is armed with four prominent teeth, which are triangular, slightly curved forwards and upwards, and flattened. The front is tridentate; the two external teeth are triangular, flattened, curved a little upwards and inwards, and small; I have seen specimens in which they are almost obsolete; the middle tooth is spiniform, and considerably longer than the others. The orbit is also furnished with similar teeth, of which there are two above, the inner one being the larger; one beneath, and one at the external angle. The surface of the carapace is convex, the regions distinctly marked, and the anterior half has several rounded elevations, but the hepatic regions are excavated towards the margin. The anterior pair of legs are of moderate size, equal; the wrist has three carinæ, each of which terminates in a small tubercle near its articulation; the hand has four distinct carinæ, two on the upper, and two on the outer surface; the moveable finger has two longitudinal grooves; and both the fingers are moderately and evenly toothed. The remaining legs are compressed and ciliated at the edges, particularly the fifth pair. The abdomen of the male has five joints, that of the female seven; the latter is of a lanceolate form, and furnished at the margin with numerous long hairs. The usual length of the carapace in English specimens, is not more than six lines, and its breadth nearly seven; but I have in my collection specimens from the Mediterranean, of which the carapace is nine-tenths of an inch in length, and an inch in breadth.

The colour in some specimens is greenish, in others purplish and brown mottled.

This must be considered as one of the least common species belonging to our coasts. It was first described by the indefatigable Montagu, who states that it was sent to him by Mr. Boys, "as the produce of the coast of Sandwich;" and he adds, "I have seen a specimen in the cabinet of Mr. Donovan, which I am assured came from the coast of Scotland." Leach mentions the latter specimen, and says that he obtained a fragment from the same locality; two other places on the south coast of Devon, Bantham and Torquay, are also named by that celebrated naturalist as its habitats. Mr. W. Thompson found three specimens washed ashore at Compton, in the Isle of Wight. The same gentleman mentions two localities in Ireland where it has been found, namely, the coast of Antrim, and Lahinch on the coast of Clare. Of its habits nothing, I believe, is known. It would appear not to approach the shore, as the only living examples on record were obtained from the refuse of trawl-fishers.

DECAPODA.
BRACHYURA.

GENUS CARCINUS, Leach.

CANCER, Auct.
CARCINUS, Leach, Edwards.

Generic Character. — *External antennæ* lodged in the inner canthus of the orbit, the basal joint narrow and sub-cylindrical. *Internal antennæ* lying obliquely in nearly circular cells. *External pedipalps* with the third joint excavated on the anterior half of the inner margin, and dilated at the outer side. *First pair of feet* somewhat unequal, the wrists with a strong spine on the inner side, standing forwards; the hands glabrous on the outer surface; *second, third, and fourth pairs* slightly compressed, with the terminal joint long, styliform, somewhat four-sided; the *fifth pair* more compressed, formed for swimming, the terminal joint lanceolate. *Carapace* slightly convex, rather broader than it is long; the front somewhat projecting, and forming, with the orbits and the latero-anterior margin, a nearly regular curve, which extends back to a line drawn through the middle of the genital region; latero-anterior margin strongly toothed. *Orbits*, oval, directed forwards, very open above, with a single fissure, both in the superior and inferior margins. *Eyes*, smaller than their peduncles. *Abdomen*, in the male, five-jointed; in the female, seven-jointed.

This genus, of which one species only is at present known, constitutes the nearest approximation amongst the swimming Crabs, to the *canceridæ;* the osculant genus in that family which bears a near affinity to this, is *Panopæus*.

DECAPODA.
BRACHYURA.

PORTUNIDÆ.

COMMON SHORE-CRAB.   HARBOUR-CRAB.

Carcinus Mænas.

Cancer Mænas,	Penn. Brit. Zool. IV. p. 3. t. iii. f. 5.
Portunus ,,	Leach, Edinb. Encycl. VII. p. 390.
Carcinus ,,	Leach, Ib. p. 429. Trans. Linn. Soc. XI. p. 314. Malac. Podophth. Brit. t. v. f. 1—4. Edw. Hist. Crust. I. p. 434.

The carapace of this common species is rather broader than it is long, minutely tuberculated, the regions very distinct and rather prominent. The front is divided into three lobes, of which the middle one is rather longer than the others; they are distinctly margined and slightly turned upwards; the orbits very open above, with a single fissure in the superior, and one in the inferior margin, and a strong tooth at the outer, and a smaller one at the inner angle. The latero-anterior margin has four strong flattened triangular teeth, directed forwards; the second and fourth more acute than the others. The latero-posterior margin extends backwards in a straight line, and the posterior margin has a distinct elevated waved border. The external antennæ are placed in a hiatus at the inner canthus of the orbit, which they do not entirely fill. The basal joint is rather narrow, and somewhat round. The internal antennæ are lodged rather obliquely in large open fossæ. The anterior pair of feet nearly equal, the wrist with a strong but not very prominent tooth at the upper and anterior angle; the hand smooth externally, the upper margin with a double longitudinal carina; the fingers toothed. The second, third, and fourth pairs slightly compressed, the terminal joint very long, styliform, somewhat four-sided; the fifth pair more compressed, the terminal joint broader and flatter than in the others, forming an approximation to the more perfectly natatory form observed in the other genera of the family. The two last joints of the second pair, and the three last of the fifth pair, ciliated on the under edge, and the latter also on the upper edge of all the joints. The abdomen, in the male, five-jointed, forming a slightly acute triangle from the base of the third joint; in the female, it is seven-jointed, broad, with rounded and ciliated margins, the terminal joint rather abruptly smaller than the preceding.

The general colour of this species is a blackish green, darker anteriorly, and often dull red underneath; they vary, however, considerably, both in the hue and in the intensity of the colour. The young are often mottled with white, and sometimes almost wholly white, with perhaps a single black spot on the centre of the carapace.

This is the only known species of the genus, and is

undoubtedly the most common Crab of our shores. On every part of the coast, it is found in numbers; on sandy beaches it is constantly left by the receding tide, concealing itself under stones, and on being disturbed, either runs to regain its natural shelter in the retiring sea, or hastily buries itself completely in the soft sand. It is, however, by no means confined to the sandy shores; it is often dredged in rather deep water, though its favourite haunt is in the former situation. Such habits as these require a power of remaining for a considerable time out of water, and we find this to be remarkably the case with this species; it cannot, it is true, like the land Crabs, live at a great distance from the sea, requiring only the moisture of a humid atmosphere, to preserve their branchiæ in a state fit for respiration, but it will remain active for many hours, and probably for days together, if it have the opportunity of burying itself in sand which is wetted with sea-water: differing in this respect from the more typical forms of the family, which require constant immersion in deep water. It will even, as Mr. Couch informs me, survive its immersion in fresh water for several hours.

This Crab is much eaten by the poorer classes on the coast, and great numbers are also brought to the London markets, the flavour being very delicate and sweet. On some parts of the coast, a small black variety is found, which the fishermen consider as a distinct species, distinguishing them as the black and the green crab. This variety is found in deeper water, and is believed to interfere with the success of their prawning, by either destroying the prawns, or frightening them away from the pots. It is certainly merely a variety.

Its food consists principally of the fry of fish, of shrimps, and other Crustacea, but it will also feed upon dead fish, and almost any other animal substance. Indeed, the most common method of taking these Crabs at Poole, where numbers are caught by the fishermen's children, is by tying a mass of the intestines of either a fowl or of any fish to a line, and hanging it over the quay: the Crabs seize upon this bait, and are drawn up in considerable numbers. Mr. Hailstone states, that they attack mussels, and that he once saw one carrying about on its hand a mussel which had closed its shell upon it. They run with considerable rapidity, and with an awkward sidelong gait; and they lurk in pools of water left by the tide, partially concealed in the sand, but with the anterior part of the carapace, including the eyes, exposed, so as to watch for the approach of their small living prey, on which they spring with great activity. They are, however, very timid and wary, and will not move if they discover that they are watched. They simulate death, if disturbed, as completely as do many coleopterous insects.

The process of exuviation takes place at various parts of the year, from spring to autumn. I have found the female carrying spawn as early as April, and as late as September.

The eggs continue to increase in size in this and in the rest of the *Portunidæ*, until the abdomen is forced backwards to an obtuse angle with the body. Like most of the Brachyura, this species buries its ova in the sand; and "when they are disengaged," says Mr. Couch, "the Crab stands high on the points of its legs, and employs a couple of them, one on each side, in working the loose tendrils to which the ova are attached." For the following interesting account of the development of this species, I am indebted to the kindness of the same indefatigable observer. "The ova come to life in about forty-eight hours or less. The following are my notes made at the time of observation on one that bred in captivity:—' It seems clear that each ovum has two investing coats, one proper to it, the other in which it is enclosed as attached to the parent. The latter has a thread, a portion of which is seen attached to the ovum after it has been thrown off. The ovum bursts on the sides opposite to this thread, and the creature first protrudes the abdominal portion, or that which is behind the carapace, and which in the ovum had been bent underneath; so that it escapes backwards. In some it appeared as if the caudal extremity protruded first; but in most it was the bent portion, and the legs were in general bent up under the thorax. They seemed, however, to find great difficulty in throwing off the loose membrane of the ovum from the thoracic portion or carapace, and almost all failed in doing this effectually, the development, perhaps, going on too rapidly, in consequence of exposure to a warm sun. I suppose, that in the natural state this is effected in the sand, by creeping backward, and thereby rubbing it off. The eyes of these young Crabs, at their first escape from the ovum, are large and sessile. In one or two instances, I thought I saw antennæ and branchiæ, or, at least, their projecting extremities; but I could not decidedly distinguish between them and the legs. The thoracic portion, or carapace, is somewhat rounded, or at least ovoid. I could see no *chelæ*, and suppose them not developed. The common legs seem bifurcate at the second joint from the extremity, and ending in a fine point; or, perhaps, the bifurcation is at the root. The abdominal and caudal portion is long and narrow, and also projecting, much resembling the corresponding portion of the Nebalia Herbstii. A considerable change or metamorphosis, must take place in these creatures before they assume their final form, thus confirming the views of Mr. J. V. Thompson on this subject; though these little Crabs differ much from the figures of the common edible Crab (*Cancer Pagurus*), as given by that gentleman.'"

This detail will be found remarkably consonant with the brief description of the Zoea of this species, by Mr. H. Goodsir,* who, however, gives figures of the more advanced development of the embryo; and it is very interesting to observe these consentaneous accounts of the interesting fact, from two observers whose investigations were carried on at a distance from each other, and without any intercommunication.

It is remarkable that this Crab, unlike the *Cancer Pagurus*, is active and pugnacious, both during the process of exuviation, and after it is completed; and although in some cases it takes place in concealment, and even, as Mr. Couch observes, whilst buried in the sand, yet they certainly appear not to require such precaution, as I have often found them running about both whilst the old crust is loosening, and in the soft state immediately subsequent to exuviation; and it is not uncommon for the males to seek the females when the latter are in this condition.

* See Jameson's Journal, xxiii. p. 181.

GENUS PORTUMNUS.

CANCER, (LATIPES),	Planci.
CANCER,	Pennant, Herbst.
PORTUMNUS,	Leach.
PLATYONYCHUS,	Latr. Edwards.

Generic Character.—*External antennæ* inserted at the inner canthus of the orbit, the basal joint small, not united to the front, moveable. *Internal antennæ* lying obliquely in their fossæ, which are but incompletely separated from the orbits. *External pedipalps* extending forwards to the antennary fossæ; the third joint elongated, emarginate at the inner margin a little behind the apex, for the articulation of the palpal portion, which is three-jointed. *Anterior feet* sub-compressed, equal; *second, third*, and *fourth pairs* with the terminal joint compressed, narrow, lanceolate, that of the first rather broader; the *fifth pair* with the penultimate joint broad, rounded, and compressed, the terminal acutely lanceolate, and broader than that of the other pairs. *Carapace*, as long as it is broad; front, narrow, toothed; latero-anterior margin arched; the posterior half of the carapace gradually narrowed; posterior margin truncate. *Orbits*, with the upper margin evenly concave, and with a single fissure, the inner canthus open. *Eyes*, not larger than their footstalks, which are rather slender and slightly curved. *Abdomen*, in the male, five-jointed, the third and fourth joints much longer than they are broad, the third being the longest; in the female, seven-jointed, less than half as broad as it is long, the second, third, and fourth joints very short, the fifth transversely quadrate, the sixth and seventh regularly diminishing to the apex.

I have adopted some characters for this genus which will imply the necessity of separating from it species which have been included by Edwards in the genus *Platyonychus* of Latreille, which is synonymous with *Portumnus* of Leach. The general form and habit of a large and very handsome species, *Platyonychus bipustulatus*, Edw., must at once strike even a casual observer as very distinct from our species, on which Dr. Leach founded his genus; and the details of many important organs will offer no less striking discrepancies. I will now venture to place before the reader some of these points in a parallel view, premising that I propose to consider our species as the type of the genus *Portumnus*, and the other as that of a distinct genus, for which I would retain Dr. Milne Edwards's name of *Platyonychus*.

PORTUMNUS.	PLATYONYCHUS.
Carapace quite as long as it is broad, with the latero-anterior margins very slightly toothed; the front tridentate. *Orbits* with a single fissure in the upper margin.	*Carapace* one fourth broader than it is long; the latero-anterior margin very strongly toothed; front quadridentate. *Orbits* with two fissures in the upper margin.
Sternum twice as long as it is broad.	*Sternum* not more than one third longer than it is broad.
Fifth pair of legs with the terminal joint broad oval, very much rounded.	*Fifth pair of legs with the terminal joint acutely lanceolate.*
Abdomen in the male five-jointed; the terminal joint not abruptly smaller than the preceding one.	*Abdomen in the male seven-jointed;* the terminal joint abruptly smaller than the preceding one.
Abdomen in the female seven-jointed; nearly three times as long as it is broad; the sides parallel, as far as the fifth joint inclusive; the terminal joint not abruptly smaller than the preceding one.	*Abdomen in the female* seven-jointed; not half as long again as it is broad; the fourth, fifth, and sixth joints forming nearly a circle, posteriorly truncated; the terminal joint only one-third the breadth of the preceding one.

Such are some of the most important characters in which these two forms differ, and on which I have thought it necessary to consider them as generically distinct. In many respects the British species more nearly resembles *Polybius Henslowii*, than it does *Platyonychus bipustulatus;* nor can I imagine, if the two in question be reduced to one generic name, how *Polybius* can consistently be considered as distinct.

It is to be remarked, that Edwards throughout quotes Leach's genus *Portumnus* as *Portunus*, from which it is of course distinct; and although it was perhaps undesirable to give to two genera so nearly allied, names so similarly spelt, yet I cannot consider this as a sufficient ground for changing the generic name from *Portumnus* to *Platyonychus*, as Latreille has done.

I have not had an opportunity of examining an American species, first described by Herbst, and afterwards by Say, and referred by Latreille and Edwards to *Platyonychus*, under the name of *Pl. ocellatus*, and therefore I am unable to state positively its relations, particularly as the abdomen has not been described by either of the naturalists who have noticed it. But I believe it will be found to belong to *Platyonychus*, as I have above restricted that genus.

DECAPODA.   *PORTUNIDÆ.*
BRACHYURA

Portumnus variegatus.

Specific character.—Front tridentate; carapace heart-shaped, not broader than it is long; terminal joint of the fifth pair of legs lanceolate.

Cancer latipes,	PENNANT, Brit. Zool. IV. t. i. f. 4. p. 5.
Portumnus variegatus,	LEACH, Edinb. Encycl. VII. p. 391. Malac. Brit. t. iv.
Platyonychus latipes,	EDWARDS, Hist. Crust. I. p. 436.

THE carapace of this species is almost evenly convex, slightly granulated, heart-shaped, as broad as it is long, the latero-anterior margins with the front almost continuously arched, the latero-posterior margins much contracted. There are four small teeth on each latero-anterior margin, exclusive of the external angle of the orbit. The front has three teeth, of which the middle one is the longest. The orbits are entire, the superior and the inferior margin regularly concave, with a strong tooth at the outer, and a smaller one at the inner angle; there is a considerable hiatus at the inner canthus, which is filled with the basal and second joints of the external antennæ. The anterior legs are of moderate length and size, rounded on the outer, and flattened on the inner sides; the wrist has a distinct carina on the superior margin, which is ciliated, and terminates anteriorly in a sharp spine; the hand is carinate above and beneath, the superior carina being, like that of the wrist, closely ciliated with short hairs, the inferior continued along the immoveable finger, which is triangular. The moveable finger is considerably curved, with a furrow on the outer side; both are obtusely toothed. The remaining legs are slightly compressed, the terminal joints of the second, third, and fourth, very narrow lanceolate; that of the fifth pair more broadly lanceolate, all acutely pointed. The abdomen of the male is long and narrow, the penultimate joint nearly quadrate, the terminal one triangular. That of the female is but little broader than that of the male, the sides parallel as far as the fifth joint inclusive, which is transversely quadrate, the penultimate and the terminal one diminishing almost regularly to the apex, which is slightly truncated.

The colour is very pale dull purplish-white, mottled with a darker hue.

Dr. Leach describes this species, with great truth, as one of the most beautiful of the British Crabs; but he is certainly in error when he calls it "the most common." It is found along the whole of the western and southern coasts; but as far as my own experience goes, and that of others of whom I have made the inquiry, not in the abundance alluded to by my lamented friend. Mr. Thompson, in giving its Irish localities, says very correctly, "It is occasionally found thrown ashore on extensive sandy beaches." It is one of the more rare and local of the Irish species. It is taken, according to Dr. Leach, by digging beneath the sand at low water mark; but there can be no doubt of its inhabiting also deep water, from the natatory character of the legs, all of which are terminated by a true swimming joint, though less strikingly so than in some of its congeners.

GENUS PORTUNUS, Leach.

CANCER,   Linn. Penn. Herbst.
PORTUNUS,   Fabr. Latr. Leach, Edwards.

Generic Character. — *External antennæ* placed in the inner canthus of the eyes, separating the orbits from the antennary fossæ, which are open in front. *External pedipalps* with the third joint quadrate, and either truncate at the inner and anterior angle, or notched at the inner margin, for the articulation of the palp. *Anterior pair of legs* generally somewhat unequal, and the wrist armed with a strong spine at the superior and interior angle; hands slightly incurved, marked with elevated lines. *The second, third, and fourth pairs of legs*, with the last joint long, styliform, slightly curved, and longitudinally grooved; *fifth pair* formed for swimming, the last and penultimate joint, being very flat, broad, and rounded. *Carapace*, rather broader than long; the latero-anterior margin four or five toothed, flattened, and thin; the front, horizontal, projecting. *Orbits*, above with two—beneath towards the outer angle, with one fissure. *Eyes*, with a short peduncle. *Abdomen*, in the male, five-jointed, triangular; in the female, seven-jointed.

The Crabs of this genus are capable of swimming with great ease, as the thin, expanded, fin-like form of the posterior feet would indicate. They are commonly termed by the fishermen, *swimming* and *flying Crab*; and, from the peculiar motion of their hinder feet, *fiddlers*. Pennant gives the name of *cleanser* Crab to one species, and the specific name *depurator*, given by Linnæus to a species of Crab presumed by Pennant and by Leach to be the one in question, would point to the same supposed office. That they do perform such an office in no very limited degree, may be concluded from the localities in which they abound, and the numbers in which they are found congregated. In the refuse of the prawn and lobster pots, where they resort for the purpose of feeding on the often half-putrid garbage which is placed there as bait, and amongst the mass of miscellaneous filth sometimes brought up by the dredge, hundreds of these *cleansers* are frequently taken.

The genus *Portunus* as established by Fabricius, was much more extensive than at present, including as it did the whole of the swimming Crabs belonging to this division; in fact, the whole family of *Portunidæ*, as far as they were then known, with the exception of *Carcinus*, which forms one of the links by which the family of *Canceridæ* are united with the present group.

DECAPODA.
BRACHYURA.

PORTUNIDÆ.

VELVET SWIMMING-CRAB.

Portunus puber.

Specific character.—Hinder feet with a longitudinal elevated line; body pubescent; front with numerous small spiny teeth.

Cancer puber,	LINN. Syst. Nat. XII. 1046.
„ *velutinus,*	PENN. Brit. Zool. IV. p. 5. t. iv. f. 8.
Portunus puber,	LEACH, Malac. Podophth. Brit. t. vi. EDW. Crust. I. p. 441.

THE carapace of this species is broader than it is long in the proportions of four to three. The anterior margin forms the segment of a circle, and each latero-anterior portion is furnished with five strong triangular teeth, the margins of all of which, excepting the last, are minutely serrated, and the points are directed somewhat forwards; the posterior is the narrowest, and finely acuminated. The orbits are very large and open, both margins minutely denticulate, with two rather deep fissures in the upper, and one in the lower; a strong denticulated tooth protects the inner canthus beneath. The eyes are round, placed on short and broad peduncles. The front is very broad, armed with a spine on each side of the centre, and a denticulated triangular tooth at the exterior extremity, between which are about three small pointed teeth. The posterior portion of the carapace is broad, the surface is granulated, and covered with a dense, short, villous coat. The first pair of legs are very robust; the spines and processes very strongly marked; on the wrist are two spines, the outer one simple and acute, the inner very strong, and furnished with two additional smaller teeth. The hand is furnished with a strong spine at the anterior and upper part, projecting over the joint of the moveable finger: the elevated portions are covered with large granulations. The fingers are longitudinally grooved, and furnished with strong irregular tubercular teeth; the points moderately acute. The second, third, and fourth pairs of legs are long, slightly grooved longitudinally, carinated above, and the terminal joint is long, slender, and pointed. The fifth pair has the last two joints much flattened; the last but one has four, and the terminal one three raised longitudinal lines, which are naked and polished: they are both furnished with a close firm fringe of hair, and the last is acuminated. The whole of the legs, as well as the carapace and thorax, are covered with a villous coat, excepting on the elevated portions, which are generally naked. The abdomen in the male forms an acute triangle, and each joint is slightly carinated transversely; in the female, it is broadly ovate.

The colours of this fine species are exceedingly bright and showy when it is alive, but soon fade after death.

Leach's figure, in his *Malacostraca Britanniæ*, is coloured after life, and exhibits a remarkable assemblage of hues, the general tint being a reddish brown, and the naked portions a bright blue.

Its velvety coat has procured for it the English name of *Velvet Crab*, and the French one of *Crâbe à laine*.

The Velvet Crab is found in considerable quantities, all along the south-western coast of England. In Cornwall and Devonshire it is very common; I have taken it in Swanage and Studland Bays, and on the southern coast of Kent, where, however, it appears to be more rare. Like some other species, it appears in much greater numbers during some seasons than in others. Mr. Hailstone has the following note respecting its occurrence at Hastings. " In July, 1834, several dozens were taken off Hastings, to the astonishment of the fishermen, who had rarely seen them here; and, since that influx, they have quite disappeared. This advance and retreat is of frequent occurrence." Mr. Embleton, in his list of the Crustacea found on the coasts of Berwickshire and North Durham, mentions its occurrence as not uncommon. Mr. Thompson records its existence on all parts of the Irish coast; and states, after Dr. Drummond, that it is taken commonly at Bangor by boys, who find it lurking under stones in rocky pools at low water. Mr. Couch observes that it is found in the adult state at a few fathoms' depth, but that the younger ones are found at low-water mark amongst stones, under which they conceal themselves. I have certainly obtained the larger specimens at a considerable distance from the shore by dredging, as well as in lobster-pots. The whole of the species of this genus are remarkably active and pugnacious; but this is, according to the testimony of Mr. Couch, " the most active and fierce of the

VELVET SWIMMING-CRAB. 93

family, running with great agility on the appearance of danger, but stopping and assuming an attitude of defence when closely pursued. It seizes an enemy suddenly, and holds him with tenacity."

It is taken with *Carcinus Mœnas*, and in the same way. I have occasionally seen it brought to the London market with that species; and it is taken in large quantities on the French coast as an article of food. It is by far the largest of the family inhabiting the European coasts, being often two inches and a half to three inches in length.

DECAPODA.
BRACHYURA.

PORTUNIDÆ.

WRINKLED SWIMMING-CRAB.

Portunus corrugatus. Leach.

Specific character.—Carapace with numerous raised serrato-granular, hairy, transverse lines; front three-lobed, the lobes crenulated, the middle one the largest; latero-anterior margin on each side five-toothed. Terminal joint of the posterior feet, with a raised median and marginal line, lanceolate and mucronate.

Cancer corrugatus,   Penn. Brit. Zool. IV. 5. t. v. f. 9. Herbst, t. vii. f. 50.
Portunus   ,,   Leach, Edinb. Encycl. VII. p. 390. Trans. Linn. Soc. XI. p. 315. Malac. Brit. t. vii. f. 1, 2. Edw. Hist. Nat. Crust. I. p. 443.

The carapace in *Portunus corrugatus,* is about four-fifths as long as it is broad, elevated, with the regions distinct, and marked with numerous transverse elevated lines; the front is three-lobed, the lobes crenulated at the margins, the middle one the largest; the latero-anterior margin five-toothed, the teeth curved, and directed forwards; the latero-posterior margin abruptly narrowed behind the posterior lateral tooth. The first pair of feet somewhat unequal, the surface rugose; the wrist with a long sharp spine at the anterior and superior angle; the hand with a sharp carina on the upper side, terminating in a sharp tooth over the joint of the finger; claws longitudinally sulcate, the superior curved, the margins furnished with numerous tubercular teeth, of which those of the larger claw are larger and irregular, those of the smaller regular and small; the second, third, and fourth pairs of feet hairy at the upper and lower edge, carinated above, and with elevated lines along the sides, the terminal joint long, slender, and styliform; the posterior feet with elevated lines on the sides of each joint, the margins of the joint ciliated, the terminal joint rather narrow, lanceolate, and mucronate. The sternum is slightly rugose. The abdomen in the male is triangular, in the female ovate; the first to the fourth joints strongly carinated transversely; the terminal joint forming an equilateral triangle.

	In.	Lines.
Length of the carapace	1	5
Breadth	1	8

The colour is reddish brown, often spotted with a brighter red.

The characters of *P. corrugatus* are so strongly marked as to preclude the possibility of its being confounded with any other species. It belongs to the same section of the genus as *P. puber*, and *P. Rondeletii*, characterised by elevated lines on the sides of the terminal and penultimate joints of the fifth pair of feet, a character which, associated as it is with a narrower form of these parts, would seem to indicate a somewhat inferior power of swimming.

It must be considered as one of the rarer species of the genus. Pennant states that it was found " on the shores of Skye, opposite to Loch Jurn." This is the first account we have of its occurrence, and Herbst's figure is copied from Pennant's. Leach mentions specimens having been taken by Mr. C. Prideaux in Plymouth Sound; and I have a fine female specimen from the same locality, given to me by my friend Dr. Miller, R.N. Mr. Couch, to whose kindness I am also indebted for a specimen, mentions it in his "Cornish Fauna" as scarce on that coast. It has been found by Dr. Johnston in Berwick Bay, but is rare. It is an Irish species, as appears from the following notice in Mr. W. Thompson's account of the Crustacea of Ireland. " The only examples of this species which I have seen, are some fine examples from Larne and Carrickfergus, in the Ordnance Collection, and a single specimen obtained on the Dublin coast, by Mr. R. Ball. Mr. J. V. Thompson notices *P. corrugatus* as inhabiting the harbour of Cove; but those so-named in his collection are the wrinkled variety of *P. depurator*." These are all the localities that I am acquainted with in which it has occurred as a British species; but it is mentioned by Edwards as being very common in the Mediterranean, although Risso does not mention it, unless his *P. Leachii* be identical with it, which is possible, as the short description given of that species agrees in every respect with *P. corrugatus*.

DECAPODA.   *PORTUNIDÆ.*
BRACHYURA.

ARCHED-FRONTED SWIMMING-CRAB.

Portunus arcuatus.   Leach.

Specific character.—Front entire, arched; latero-anterior margin five-toothed; the penultimate tooth the smallest.

Portunus arcuatus,	Leach, Malac. Brit. t. vii. f. 5, 6. W. Thompson, Ann. and Mag. Nat. Hist. X. p. 283.
„ *Rondeletii.*	Risso, Hist. Nat. des Crust. de Nice, t. i. f. 3. Id. Hist. Nat. de l'Eur. Mérid. V. p. 2. Edw. Hist. Nat. Crust. I. p. 444.
Var. fronte emarginato.	
Portunus emarginatus,	Leach, l. c. vii. f. 3, 4.

I HAVE followed Milne Edwards in adopting the suggestion of Leach, that the *Portunus emarginatus* of the latter is only a variety of his *P. arcuatus.* In retaining

Leach's name for the species, in which I have been preceded by Mr. W. Thompson, of Belfast, I believe that I follow the strict law of priority; as the early parts of the "Malacostraca Britanniæ" were published in 1815, and Risso's "Histoire Naturelle des Crustacés des environs de Nice" not until the following year. The specific name *Rondeletii* is retained by Risso in his Natural History of Southern Europe, which was published in 1826, so that he was either not aware of Leach's figure, or not satisfied of its specific identity with his own species. Milne Edwards has also kept Risso's name against the law of priority of description.

The carapace is four-fifths as long as it is broad; considerably raised, the regions distinct, the surface granulated; the anterior portion slightly scabrous; the anterior margin describing nearly a semicircular arch, of which the front forms a continuous portion; latero-anterior margin on each side armed with five teeth, including that at the external canthus of the eye; the fourth being the smallest, and the fifth prominent and acute. Front entire, except in the variety named by Dr. Leach *emarginatus*, in which it is slightly excavated, the margin granulated and fringed with rather long hair; the posterior portion of the carapace broad, the posterior margin nearly straight; orbits with two fissures on the upper margin, and one beneath. The anterior feet in the male very robust; the wrist armed with a single prominent and acute spine; the hand with a double carina on the superior edge, each terminating in a small tubercle; the fingers strong, armed with numerous tuberculous teeth, and each having two carinæ on the outer surface, and a carina and a groove on the inner; the remaining pairs of feet rather slender; the third and fourth the longest, and the second shorter; the fifth pair fringed with long hair; the terminal joint lanceolate, very acute. Abdomen in the male regularly triangular; in the female semi-ovate, slightly carinated, the terminal articulation triangular. The colour of this species is a dull blackish-brown above, paler beneath, and with a tinge of red; the legs paler than the body.

The habits of this species are very similar to those of the other species of the genus as far as they have hitherto been observed. They are active, bold, swimming with agility, and seizing with great sharpness, and pinching severely with their acute claws. They are gregarious, like most of their congeners; and I found them extremely abundant at Bognor, where they constantly infest the prawn-pots and, as the fishermen believe, keep the prawns from the bait.

I believe this species will prove, upon further observation, to be more generally distributed than has hitherto been supposed. Dr. Leach gives the more northern coasts of England as its usual habitat; I have dredged it in Poole Harbour, and in the neighbouring bays of Studland and Swanage, and plentifully at Bognor. Mr. Eyton sent me specimens from the Welsh coast. Mr. Couch does not, however, give it a place in his " Cornish Fauna;" nor does it occur in the late Mr. Hailstone's MS. notes of Crustacea taken at Hastings. In Ireland, Mr. W. Thompson has taken it " when dredging in deep water in the loughs of Strangford and Belfast;" and he adds, "it was procured by our party when dredging in Killery and Roundstone Bays on the Western coast." Mr. Ball also found it cast on shore at Portmarnock.

I have never found the variety named by Leach *P. emarginatus*. Of the hundreds which I have taken, all possessed the arched and entire front assigned by him to his *P. arcuatus*. The original specimen of his *emarginatus* is in the British Museum, and a figure of it is given below.

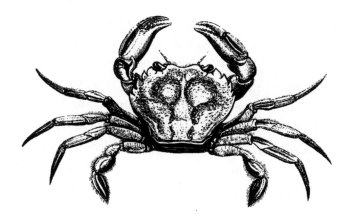

DECAPODA.　　　　　　　　　　　　　　　　　　　　PORTUNIDÆ.
BRACHYURA.

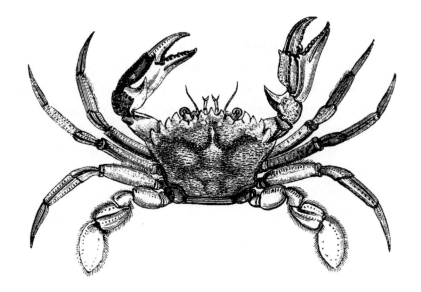

CLEANSER SWIMMING-CRAB.

Portunus depurator. Leach.

Specific character.—Front armed with three triangular teeth, and a small one on each side, over the inner angle of the orbit; latero-anterior margin with five teeth; carapace irregularly granulate, scabrous. Terminal joint of the posterior feet broadly oval, smooth.

Cancer depurator, (?)	LINN. Syst. Nat. XII. 1043, 23.
,, ,, var.	PENN. Brit. Zool. IV. t iv. fig. 6. A.
Portunus depurator,	LEACH. Edinb. Encycl. VII. p. 390. Trans. Linn. Soc. XI. p. 317. Malac. Brit. t. ix. f. 1, 2.
,, *plicatus*,	RISSO, Crust. de Nice. Id. Hist. Nat. de l'Eur· Mérid. V. p. 3. EDW. Hist. Crust, I. p. 442.

THE carapace of this species is very uneven on the surface, the regions being distinctly marked, and all the elevated parts scabrous, with unequal raised granules or points, some round, others elongated. The latero-anterior margin on each side armed with five triangular teeth, slightly curved forwards and sharp pointed. The front has three projecting flat teeth, of which the middle one is rather the longest, and a smaller one at the outer side, and a little posterior to these, over the inner angle of the orbit. The orbits are large, opening forwards and upwards; the eyes large and the peduncles very short. First pair of legs slightly unequal, elegantly sculptured; the wrist having the superior area granular, bounded by raised lines, of which the outer one is furnished with two or three small teeth, and the inner terminates anteriorly in a sharp spine; the hand has five longitudinal raised lines, which are granular, or slightly denticulate, and the superior one terminates in a small sharp spine over the joint of the finger; the claws are longitudinally carinated, and furnished with very distinct rounded tubercles. The second, third, and fourth legs are long and slender, with a double carina running along the superior edge, the terminal joint very long, slender, and sharp pointed. The fifth pair very much flattened, the joints ciliated at the margin, and sculptured, excepting the terminal one, which is flat, smooth, and oval. The abdomen in both sexes has the second and third joints acutely carinated transversely. That of the male is triangular; that of the female very broad and ciliated with long hairs; the third to the sixth joints broader than the first two, the seventh abruptly narrower.

The colour is generally a pale reddish brown; in the younger ones flesh-coloured.

The sculpture in this species varies greatly in degree. The specimen figured in Leach's Malacostraca, and which may be considered as a fair representation of the ordinary appearance of the adult individual, is comparatively smooth; whilst a younger one, which I have from the Mediterranean, is very sharply and elegantly sculptured. There is, in fact, no species of the genus, and scarcely any of the whole order, the surface of which is more minutely and beautifully relieved, and this is particularly the case with the hands and wrists, the inequalities of which are most delicately picked out.

The early synonymy assigned to this species by Leach is, to say the least, exceedingly doubtful. The figures to which Linnæus refers in his synonymes of *Cancer depurator*, may be referred to two or three other species, with quite as great probability as to this. But as Fabricius and Leach have both appropriated the specific name of *depurator* to the species, and as there is no proof whatever that it was originally given to another species, I have preferred retaining it, to the adoption of the name of *plicatus*, subsequently assigned to it by Risso, and continued by Edwards.

This is not an unfrequent species on our coasts. In the north it has been recorded by Mr. Embleton as occasionally brought from deep water in Embleton Bay, adhering to the nets of the fishermen. Leach states that it is the most common of all the species of the genus; but like many others it is local, although, like them, very numerous where it does occur. This is confirmed by the observation of Mr. Ball, quoted by Mr. Thompson in his account of the Crustacea of Ireland. " We have," says the latter gentleman, " dredged it in Strangford Lough, in the open sea, off Down, and on the Connaught coast. During some weeks spent at Bangor, near the entrance of Belfast Bay, in the autumn of 1835, I found this to be the most common species of Crab thrown by the waves upon the beach. Mr. R. Ball mentions that the *P. depurator* is local, but abundant where it does occur about Youghal." I have dredged it in Studland Bay, in Dorsetshire; but have not found it on the coasts of Sussex and Kent, where I have found other species in great plenty. Mr. Hailstone, however, states that it is frequently caught at Hastings in the shrimping-net.

The habits of this species are doubtless similar to those of the rest of the genus. I am not aware of the period of its spawning in this country, but Risso states that it occurs in March and December in the Mediterranean.

DECAPODA.   *PORTUNIDÆ.*
BRACHYURA.

MARBLED SWIMMING-CRAB.

Portunus marmoreus. Leach.

Specific character.—Carapace even, very slightly granulated, without hairs; latero-anterior margin armed with five teeth on each side; front three-toothed, the teeth rather obtuse, the middle one the longest; hands with four carinæ, slightly denticulate; terminal joint of the posterior feet without raised lines, the apex mucronate.

Portunus marmoreus,   Leach, Malac. Brit. t. viii. Edw. Hist. Crust. I. p. 442.

The general form and the whole of the characters of this elegant species resemble so exceedingly that of *P. holsatus*, that I am almost imperatively forced to consider them as varieties of one species The carapace is somewhat convex, with the regions moderately distinct; the surface obsoletely minutely granulated, smooth and naked, with an arched line of very slightly raised points, separating the hepatic from the branchial regions, and a sulcus between the latter and the genital. Latero-anterior margin with five acute flattened triangular teeth: the points directed forwards, the last being the most acute and the longest. Posterior margin waved, broad, moderately hollowed at each side. Front with three teeth, the middle one slightly longer than the others—all rather obtuse. Anterior feet strong, angular; the wrist with a rough irregularly rhomboid area on the upper surface, bounded by a raised denticulate line; the anterior angle with a very strong tooth. Hands with four distinct carinæ, which are generally slightly denticulate; the superior one terminating in a small sharp tooth. Fingers longitudinally carinated, strongly tuberculated; the moveable one much curved. The second, third, and fourth pairs of feet rather slender, compressed; the terminal joint curved, hairy on the inferior edge; the fifth pair having no raised lines on the terminal and penultimate joints; the whole fringed with hair; the terminal joint very smooth, ovate and slightly mucronate. Abdomen in the male, five-jointed, triangular; the second and third joints transversely acutely carinated; in the female seven-jointed, also triangular, but broader, and with the second and third joints similarly carinated.

	In.	Lines.
Length of the carapace	1	3
Breadth of ditto	1	6

The colours of this species are exceedingly varied and beautiful, particularly in the males. Buff, light-brown, deeper brown, and brownish red are arranged over the carapace, in varied but always exactly symmetrical patterns. The only way in which these beautiful markings can be preserved is, by raising the carapace, taking out the soft parts and drying the specimens in a shady place in a brisk current of air. If they are put into spirit, the whole of the beauty of the colour is lost.

The younger specimens do not possess these markings. They are, as Dr. Leach has observed, of a plain brown colour, and much resemble the fry of *Portunus depurator*, from which they may be easily separated by their more considerable convexity. It must be considered as one of the more local species of the genus, occurring, however, in considerable numbers in its favourite localities. It was first discovered by Montagu, who sent specimens to Dr. Leach for description; and who appears, from Leach's quotation, to have named it, "Cancer *pinnatus* marmoreus." It is not uncommon, according to the latter author, " on the sandy shores of the southern coast of Devon, from Torcross to the mouth of the river Ex, and is frequently found entangled in the shore-nets of the fishermen, or thrown on the shore after heavy gales of wind. It is included in Mr. Couch's "Cornish Fauna," but without any remark. It does not appear to have been hitherto taken on the coast of Ireland; and Mr. H. Goodsir mentions it as not common as a Scottish species. At Hastings, I procured a single specimen, which I found in a shop where shells, crustacea, and other marine productions were sold, but it was certainly native at that place; and at Sandgate, in the month of May, 1844, I procured by dredging nearly four hundred specimens at two casts of the dredge, of which about three-fourths were females. Several of these were carrying spawn, which is of a rich orange colour.

It is very curious to observe how local these "cleansers" are. In the former year, at Bognor, I found multitudes of *Portunus Rondeletii*, which absolutely swarmed in the prawn and lobster-pots, but not a specimen of any other species was obtained there. The place of these is supplied at Sandgate by the present species, whilst farther to the west, *P. puber*, and *P. depurator* appear to occupy the ground and perform the same important office of scavengers of the sea.

There is another fact relative to this species which is worth recording, and that is, the extent to which they are infested with a remarkable parasite, occupying the space between the folded abdomen and the sternum, and having the *primâ facie* appearance of a bag of immature eggs. Both males and females are equally obnoxious to it; and from its size and situation it must present an insuperable barrier to impregnation. It consists principally of a mass of minute eggs, which are arranged in bundles attached to filaments, like bunches of grapes; the alimentary canal passes directly through the body, the mouth being attached to the intestine of the crab, which it pierces near its extremity, and from which, in all probability, it derives its nourishment. The anal opening, which is distinct and obvious, is visible without removing the parasite from its position. The whole is of a rounded trihedral form, and is covered by a tough but thin integument. I have occasionally found it infesting *Carcinus mœnas*, but never in such numbers as on the present species.

DECAPODA.   *PORTUNIDÆ.*
  BRACHYURA.

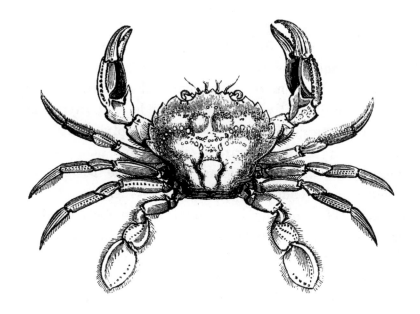

LIVID SWIMMING-CRAB.

Portunus holsatus.   Fabr.

Specific character.—Carapace somewhat depressed, minutely granulated; latero-anterior margin with five strong flattened teeth; front with three nearly equal teeth; posterior margin very largely emarginate at the angles for the hinder feet; hands with denticulate carinæ; last joint of the fifth pair roundly oval, more than half as broad as long.

Portunus holsatus,	Fabr. Suppl. p. 336. (Edw.)	Edw. Hist. Nat. Crust. I. p. 442.
„ *lividus,*	Leach, Malac. Brit. pl. ix. figs. 3. 4.	

It is extremely difficult to assign any very satisfactory distinctive character to this species. Its great resemblance to *P. marmoreus,*—at least to all the specimens which I have in my possession, fully justify the belief that they may be merely varieties; although there are certain comparative characters which, as they are pretty constant, render it necessary that further investigations should be made before their identity can be fully established. Then the whole contour of the animal is more strongly marked in the present species; the marginal teeth are more prominent; the margins of the orbit more distinctly granulated; the latero-posterior margin much more contracted and more deeply emarginate at the angles; the outer carina of the hand, more strongly denticulate; and the terminal joint of the posterior feet rounder and broader in proportion to its length. In other respects the similarity is so great in the form of all the parts, as fully to justify Dr. Milne Edwards's remark of their "extreme resemblance." It is matter of surprise that Dr. Leach should not have observed this close relation of these two species; but that he should, on the contrary, have stated that *P. lividus* [*holsatus*] most nearly resembles *P. depurator*, a species from which, in fact, it differs most obviously. It is remarkable that the specimens of *P. marmoreus* in the British Museum, which were collected by Dr. Leach, differ much more from *holsatus*, than those which I have myself procured; the hand having in all those unarmed carinæ, and the upper margin of the orbit without granulations. The figures in Dr. Leach's beautiful work, also magnify the distinctions far beyond the truth.

The occurrence of this Crab is extremely rare on our coasts; Dr. Leach mentions his having found a single specimen amongst a number of *P. depurator* that were taken in the Frith of Forth at Newhaven, and that he observed another in the collection of Montagu; but there is a fine series in the British collection of the British Museum, which must have been procured after the " Malacostraca

Britanniæ" was published. It is not mentioned by Mr. H. Goodsir as occurring within his notice on the Scottish coast; nor does Mr. Couch give any account of its occurrence in Cornwall. In Ireland, however, according to Mr. W. Thompson's statement, it has occurred repeatedly; but as it appears to me that faded specimens of *P. marmoreus* might be easily mistaken for this species, it is always desirable that they should be compared with those well distinguished specimens which exist in the British Museum. The following is Mr. Thompson's notice to which I have referred. "Templeton mentions it as found by him 'on the shore at Dunfanaghy.' We have dredged it on more than one occasion in Belfast Bay, and have obtained it on the beach of Carnlough, county of Antrim. In Mr. R. Ball's collection, are several specimens which were dredged in Dublin Bay." It is mentioned by Milne Edwards as occurring on the French coast.

DECAPODA. *PORTUNIDÆ.*
BRACHYURA.

DWARF SWIMMING-CRAB.

Portunus pusillus.

Specific character.—Carapace considerably raised, rugose; front three-lobed, much advanced; latero-anterior margin with five teeth.

Portunus pusillus,	LEACH, Malac. Brit. t. ix. f. 5—8. EDWARDS, Nat. Hist. Crust. I. p. 444.
,, *maculatus,*	RISSO, Hist. Nat. Eu. Mérid. v. p. 5. ROUX, Crust. Mediter. t. xxxi.

THE carapace of this species is broader than it is long, considerably elevated, and with the regions remarkably distinct; the surface is rugose, and irregularly granulated. The front is advanced much beyond the orbits, flattened, and three-lobed, the middle lobe being longer than the others: the latero-anterior margin has five teeth, (including the outer angle of the orbit,) of which the posterior one is the most acute, and the most curved. The posterior margin is almost straight. The first pair of legs are large and robust; the wrist is armed with a very strong spine on the inner and anterior angle; the hand has a double carina above; the fingers are strongly tuberculated, and the moveable one has a shallow longitudinal groove on the upper and outer margin. The second, third, and fourth pairs are slightly compressed and grooved. The fifth pair has the penultimate joint grooved, and the terminal joint is oval; they are both ciliated all round.

The abdomen in the male is broadest at the base of the third joint, the remainder forming a regular acute angled triangle; that of the female is ovate-lanceolate and ciliated at the margin.

The colour is reddish-brown, often with red spots on the back. In some specimens the colour is lighter, being of a pale red with darker spots. The legs are usually annulated with similar colours.

This very pretty species was first described by Dr. Leach in the eleventh volume of the Transactions of the Linnean Society, under its present name. Subsequently to this, Risso described it in his Natural History of Southern Europe, giving it the name of *P. maculatus*, which Roux very improperly retained in his Crustacés de la Méditerranée, notwithstanding he was aware of the priority of Leach's name. It inhabits deep water, and is common on the coast of Devonshire and Cornwall; it occurs all along the southern coast, and is also found in the Frith of Forth, and I have specimens taken by Mr. McAndrew off the Isle of Man. On its occurrence as an Irish species, Mr. Thompson has the following remarks, " It is ordinarily taken by us when dredging in the loughs of Strangford and Belfast. At the Killeries in Connemara, it has similarly occurred, as well as in Dublin Bay. In the South, too, it has been taken in the harbour of Cove. I have several times taken it in the stomach of fishes; in one instance, in a *Trigla*

Gurnardus, taken in the open sea off Dover." It is found also in the Mediterranean, and off the coast of France. It spawns in June, and the eggs are of a reddish orange colour.

Its ordinary size is about four lines in length; this is the size of the figures of Roux, and of those of Leach; but it occasionally grows much larger, as one of the specimens, a male, taken by Mr. Mc Andrew off the Isle of Man, is fully an inch in breadth, by eight-tenths in length.

GENUS POLYBIUS. Leach.

POLYBIUS.   Leach, Edwards.
PLATYONICHUS.   Latr.

Generic character. — *External antennæ* with the basal joint round, detached, moveable, with the remaining portion lodged in a hiatus at the inner canthus of the orbit, which it does not fill. *Internal antennæ* in fossæ, which are entirely open forwards. *Eyes* larger than their peduncles, which are short. *External pedipalps* with the third joint subquadrate, longer than broad, and slightly notched at its inner margin, near the anterior angle. *Carapace* nearly orbicular, slightly contracted posteriorly. *Anterior pair of legs* equal, the pincers curved. *Second, third, and fourth pairs* compressed, the terminal joint flattened, thin, broad, and lanceolate. The *fifth pair* with the penultimate joint much flattened; the terminal one very large, oval, foliaceous. *Carapace* much depressed, the anterior margin semicircular. *Orbits* with two fissures in the superior, and one in the inferior margin; a hiatus at the inner angle, and a small tooth at the outer. *Abdomen* of the male, five-jointed, the first, second, and third joints very short and broad, and transversely carinated; of the female, seven-jointed, the sides nearly parallel as far as the middle of the sixth joint.

The structure of this genus, of which a single species only is known, is of a more decidedly natatory character than any other brachyurous form found on our shores. It is on this account that it has been with great propriety considered as generically distinct from *Portumnus*, with which, however, it stands in very near relation.

DECAPODA.
BRACHYURA.

HENSLOW'S SWIMMING-CRAB.

Polybius Henslowii. Leach.

Polybius Henslowii, Leach, Malac. Brit. t. ix. B. Edwards. Crust. I. p. 439.

This species, the only one of the genus at present known, exhibits the natatory structure to the greatest extent of any of the British examples of this family. The carapace is remarkably flat, even in the female, and the regions are very indistinctly marked; it is all over minutely granulated. Its form is nearly orbicular; the latero-anterior margins, with the orbits and front, forming a semicircle, and the latero-posterior margins being but little contracted: the front is flat, and has five teeth, the external of which on each side belongs to the orbit: the latero-anterior margin has five flat teeth, the points directed somewhat forwards.

The first pair of legs are nearly equal: the wrist has two sharp teeth on the anterior margin, of which the inner is much the more prominent, and a third tooth is found at the outer and anterior angle, which forms the commencement of a carina, which extends the whole length of the wrist. The hand is compressed, and has three low but sharp longitudinal carinæ, the spaces between them being slightly hollowed: the fingers are much compressed, somewhat incurved, as long as the hand. The three following pairs are much compressed, particularly the last two joints; the terminal one being very thin and lanceolate. The last four joints are ciliated on the inferior margin. The fifth pair have the last two joints very broad and flat; the penultimate being irregularly quadrate, and the terminal one broadly oval, slightly acuminated at the apex. The abdomen in the male consists of five joints, of which the first, second, and the base of the third are transversely carinated; the third joint is broadest at the base, and becomes moderately contracted with a slight notch; the fifth is rather acutely triangular. In the female, the abdomen is seven-jointed; the first three joints transversely carinated; the fifth joint suddenly smaller than the preceding one, and obtusely triangular.

The colour is a rich reddish-brown, which becomes a pale salmon-colour in drying. The under parts are pale.

Of this species, which is very local in its distribution, and probably nowhere existing in great numbers, there is a specimen in the Banksian collection in the Linnean Society, which was taken on the coast of Spain. It was first discovered on our shores by Professor Henslow in a herring-net, on the north coast of Devon, in 1817, and by him communicated to Dr. Leach, who named the species after its discoverer, assigning to it also a new generic appellation. It was afterwards found by Mr. Prideaux on the south-western coast of Devon; also in herring-nets on the Dorsetshire coast, amongst the refuse of the nets of fishermen, by the late Rev. Dr. Goodall. I have also obtained it at Hastings, and received it, by the kindness of Mr. Couch, from Cornwall, and by my friend Mr. Dixon, from Worthing.

The following observations on the habits of this species are from the Cornish Fauna of Mr. Couch; and as this gentleman appears to be the only one who has ever observed its habits, I make no apology for quoting his account entire. " This is, more than any others, a swimming-crab; for whilst the other British species of this family are only able to shoot themselves from one low prominence to another, the Nipper Crab, as our fishermen term it, mounts to the surface over the deepest water, in pursuit of its prey; among which are numbered the most active fishes, as the Mackerel and the running Pollock; the skin of which it pierces with its sharp pincers, keeping its hold until its terrified victim becomes exhausted. We are witnesses of this curious method of obtaining food in the summer only, at which time the fishermen's nets intercept them and their prey together; and it is probable, that in colder weather, they keep at the bottom in deep water, from which, however, I have never seen them brought in the stomachs of fishes. So far as my observation extends, it is chiefly or only the male that pursues this actively predaceous existence; but that for a time they also remain quiet, as appears from the fact that while for the most part the smooth and flattened carapace is clean, I have seen it covered with small corallines."*

This interesting narrative is perfectly consistent with the remarkable natatory structure of the species, evinced in the form of the carapace and the structure of the legs, and with the sharpness and strength of the claws.

* Couch's Cornish Fauna, p. 71.

GENUS PINNOTHERES, Latr.

| Cancer, | Linn. Fabr. Herbst, Penn. |
| Pinnotheres, | Latr. Leach, Edwards. |

Generic character.—*External antennæ* very short, occupying the inner canthus of the orbit. *External pedipalps* oblique; the second articulation rudimentary, the third large, and forming the whole valvular portion; the fourth inserted at the extremity of the previous one; and the fifth giving attachment to the sixth at the middle of its anterior margin, resembling the thumb of a didactyle hand. *Anterior legs* equal, the remaining pairs somewhat compressed; the terminal joint acute, curved, and strong. *Eyes* inserted on very short peduncles, distant. *Orbits* nearly circular. *Carapace* nearly circular, rounded at the anterior margin. *Front* not united to the epistome. *Abdomen* seven-jointed in both sexes; that of the male small, of the female extremely broad, round, and prominent.

The species of this genus are very remarkable from the peculiarity of their being indebted to animals of a very different class for protection, although not truly parasitic. They are found always to inhabit the shells of the Bivalve Mollusca, principally of the genera *Mytilus, Modiolus,* and *Pinna,* and occasionally also of *Ostrea, Cardium,* and other genera; and this habit, which was well known to the ancients, gave rise to some interesting and curious hypotheses and fables, which will be alluded to hereafter. The males are always very much smaller than the females, and the crust of the former is as hard as in other brachyurous forms; but the female is comparatively very large, almost globular, and remarkably soft; the latter character being doubtless the cause of its requiring the efficient protection of the shells of Mollusca. In other allied forms a somewhat analogous habit is observed; the soft body of *Elamene* and *Hymenosoma* demanding extrinsic protection, which they obtain by appropriating to themselves small single shells of dead acephalous Mollusca, as I have myself seen in several instances,—a fact which affords a collateral argument in favour of Milne Edwards's association of these different genera in one family.

The species of the present genus even yet require careful revision; and I have found it necessary to comprehend the whole of Dr. Leach's six species in two,—which, however, I have not done without the most deliberate consideration.

DECAPODA. PINNOTHERIDÆ.
BRACHYURA.

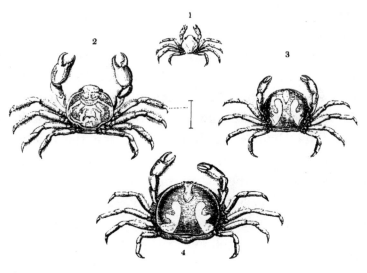

1 Pinnotheres Latreillii. Leach.   2 Pinnotheres varians. Leach.
3 Pinnotheres Pisum. Leach.   4 Pinnotheres Cranchii. Leach.

COMMON PEA-CRAB.

Pinnotheres Pisum.

Specific character.—Front of the male projecting; carapace of the female uniformly rounded at the anterior margin; abdomen in the latter sex broader than it is long.

Cancer Pisum,	PENNANT, Brit. Zool. IV. t. i. f. i. p. i. HERBST, I. p. 95, t. 2, f. 21. FABR. Suppl. Ent. 343.
Pinnotheres Pisum,	LATR. Hist. Nat. des Crust. VI. p. 83. LEACH, Mal. Brit. t. xiv. f. 2, 3, (fem.) EDWARDS, Hist. Nat. des Crust. II. p. 31.
,, *Cranchii,*	LEACH, l. c. fig. 4, 5, (fem.)
,, *Latreillii,*	LEACH, l. c. f. 6, 7, 8, (mas immat. ?)
,, *varians,*	LEACH, l. c. f. 9, 10, 11. (mas.)

THE sexes in all the species of this genus differ so remarkably, that a separate description is necessary.

MALE. (Figs. 1 and 2.) The carapace is nearly orbicular, very slightly narrowed forwards, convex, glabrous, and solid; the front projecting, arched, and entire; the latero-posterior margin slightly hollowed. The eyes small, round, and filling the orbits. Sternum large and orbicular. Anterior feet robust, the hands large, ovate, with two lines of hairs beneath; the fingers much curved, the moveable one with a single tooth. The remaining pairs of legs fringed with hair both above and below, terminating in a hooked claw. The abdomen is broadest at the third joint, becomes narrower from this to the fifth, the sixth is a very little broader, and the last abruptly narrower.

FEMALE. (Figs. 3 and 4.) The carapace in this sex is nearly orbicular, rather broader than it is long, without any projecting front, or hollows at the latero-posterior margin, soft and glabrous. The hands are oblong, weak, and furnished beneath with a single line of hairs. The remaining legs slender, the thighs fringed with a line of hairs on the upper side only. Abdomen very large, broader than it is long, almost evenly rounded.

The colour of the male varies; it is usually of a pale yellowish grey, with rather darker symmetrical markings. The female is ordinarily slightly transparent, brown above, a yellow spot over the front, and an irregular one on each branchial region; the abdomen yellow, with a central large triangular brown spot extending from the base nearly to the extremity.

In accordance with the opinion of Mr. W. Thompson, I cannot but believe that the individual figured by Leach under the name of *P. Latreillii*, which he considered as an immature female, in which he is followed by Milne Edwards, is in fact a young male. The form and apparent consistence of the carapace, the form of the hands, and the colour, are all in favour of such an opinion. The form of the abdomen is not at all at variance with it, as in many species this part is very similar in the young of the two sexes.

It is very remarkable that Leach should have failed to detect the male and female of this very common species as being specifically identical. They are frequently found together, and yet he describes the female as one species, *P. Pisum*, avowing his ignorance of the male, and the male as another, *P. varians*, acknowledging himself similarly unacquainted with the female, "unless she be *P. Pisum*." After a careful examination of the subject, I have come to the conclusion that the first four species of Leach are all to be referred to one; an opinion in conformation with that of Dr. Edwards.

This species of Pinnotheres is very commonly found in the common mussel, *Mytilus edulis*, on many parts of our coast; and especially in those which are found in rather deep water. On one occasion I dredged great numbers of these Mollusca on the coast of Dorset, and found by far the greater number of them with one or two of these little soft-bodied crabs within their shells; for the females are much more common than the males. The latter sex I have occasionally taken apart from the mussel-shells, the former never. They also inhabit the shells of *Modiolus vulgaris*, and occasionally also the common cockle, *Cardium edule*, in which I have now and then found them, as well as very rarely in the oyster, in which Mr. Ball also states that he has taken them. The following account of some circumstances respecting this crab is extracted from my friend Mr. W. Thompson's observations on the Crustacea of Ireland,* and is too interesting to admit of being curtailed.

* "Ann. and Mag. Nat. Hist." vol. x. p. 284.

"The smallest *Pinnotheres* I have seen was found by Mr. Hyndman, in a living *Cardium exiguum*, dredged by us in Strangford Lough in October, 1834. It is a male; the carapace is under a line in length; the entire breadth of the crab from the extremities of the outstretched legs is three lines. The cardium is under three lines in length, and barely exceeds that admeasurement in breadth; so that the crab when in the position just mentioned must have, on both sides, touched the walls of its chosen prison. The *Pinnotheres* likewise inhabits the *Cardium edule*. Before me is one of these crabs, of which the carapace is two lines in breadth, obtained by Mr. Hyndman in a full grown *C. edule* from Strangford Lough; but from the Sligo coast, where this crab attains an extraordinary large size, a crab with a carapace four lines in breadth, and with outstretched legs seven lines across, was once kindly brought to me by Lord Enniskillen. Mr. R. Ball informs me that on two occasions he obtained a great number of the Pinnotheres, and which were all males, from the *Cardium edule* taken at Youghal,—about nine out of every ten cockles contained a crab. On opening oysters in Tenby, in Wales, he has likewise procured the *Pinnotheres*. This crab, like the *Pagurus*, occupies different species of shells according to its size, and at every age generally selects such as with outstretched legs it would fill from side to side."

It is a point of considerable interest as connected with this species, that it formed one of the subjects of Mr. Vaughan Thompson's investigations on the transformations of Crustacea, and the description with figures of the Zoëa of *Pinnotheres* as given in a paper by that gentleman in the "Entomological Magazine."*

* Vol. iii. p. 85.

"As the females are found with an amazing group of ova under their abdominal plate," says this author, "in spring, summer, and autumn, it is probable that they have several successive broods. This circumstance renders it no difficult matter to select a number of females with mature ova at any convenient time, and to preserve them alive in sea water for a few days, or until the ova should hatch.

"From several females selected and kept alive after the above manner, I had the satisfaction to see the ova hatch in great numbers, under the form of a new kind of *Zoë*, differing from all those previously discovered, with the front and lateral spines deflected, so as to resemble a tripod. In this stage the minute animals are all like the *Zoea*, purely natatory, disperse themselves abroad, probably undergo a further change, and may be supposed to gain an easy access within the bivalve shells, before they lose their power of swimming."

I add a copy of Mr. Thompson's figures of this interesting state of the animal, the accuracy of which I can attest from my own observation.

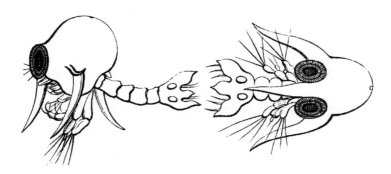

DECAPODA.
BRACHYURA.

PINNOTHERIDÆ.

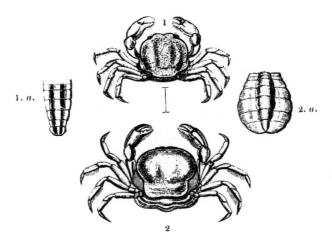

Pinnotheres veterum. Leach. Male and Female.

PINNA PEA-CRAB.

Pinnotheres veterum.

Specific character.—*Male.* Carapace subquadrate, rounded, the front slightly emarginate. *Female.*— Carapace broader than it is long; abdomen broadly ovate, longer than it is broad.

Pinnotheres veterum,		Bosc, Hist. Nat. des Crust. I. 243. LEACH, Malac. Brit. t. xv. f. 1, 2, 3, 4, 5. EDW. Hist. des Crust. II. p. 32.
„	*Pinnæ,*	LEACH, Edinb. Encycl. VII. p. 431. V. THOMPSON, Ent. Mag. III. p. 89.
„	*Montagui,*	LEACH, Malac. Brit. t. xv. f. 6. EDW. l. c. p. 32.

THE male of this species has the carapace less solid than *P. Pisum*, rather broader than it is long, rounded, slightly quadrate, with the front slightly emarginate; the hands are ovate, with the fingers arched; the remaining feet very similar to those of *P. Pisum.* The abdomen gradually and evenly decreasing towards the extremities, the last joint evenly rounded, nearly semicircular. In the variety termed *P. Montagui* by Dr. Leach, this joint is abruptly broader. In the female the carapace is rounded, broader than it is long, very minutely punctulate; the front transverse, slightly arched, scarcely emarginate at the middle. "The anterior feet with a small spine on the inferior margin of the hand." The abdomen is evenly ovate, broadest at the fourth and fifth joints, broadly carinate along the middle, the last joint emarginate.

Colour in both sexes almost uniformly brown.

This species differs sufficiently from the former, in either sex, to be distinguished at the first glance. Its habits, however, are perfectly similar, as far as we have an opportunity of knowing them, but it is much less common than the other on our coasts. It was first discovered to be an English species by the indefatigable Montagu, who found both sexes in *Pinnæ* from the Salcombe Estuary in Devonshire; and it was subsequently taken by Cranch in the same locality. Vaughan Thompson records its being found on the Irish coast, "both in *Pinnæ* and in *Modioli*." It has not, as far as I am informed, been found on any other part of the English coast but that already mentioned, nor has it yet been taken in Scotland.

Its favourite haunt justifies the name which Leach first assigned to it, *P. Pinnæ;* although he afterwards very properly adopted the name previously given to it by Bosc. It is found in the *Pinna ingens*, both on our coast and in the Mediterranean; it has also been taken in *Modioli*, and in the common oyster. There can be no doubt that it was of this species that the ancients, aware of its peculiar mode of existence, formed such absurd notions. It is not, indeed, wonderful that with such imperfect ideas of the value and bearing of natural phenomena, and with a love of the marvellous, which no Baconian philosophy then existed to correct, the relations of these little interesting parasites to their gigantic hosts should have given rise to legends as amusing as they were false; and we find that Cicero and Pliny and Oppian have, in various degrees, given currency to the most erroneous notions. Aristotle, indeed, with his accustomed accuracy, first, and alone amongst the ancients, offered any correct ideas of their habits; but even he states that the life of the protecting shell-fish depends for its continuance on that of its little guest. The absurdities of the other ancient authors whom I have named, are only worthy of recital as examples of the danger of trusting to the assertions and conclusions of those who have no general principles to guide them,—a danger not even in the present enlightened age, altogether to be neglected as chimerical.

I have thought it necessary, on the most mature consideration, to merge *Pinnotheres Montagui* of Leach as a synonyme of this species,—a result to which I am led by a careful examination of the single specimen on which that species was founded, and which is in the British Museum. The sole appreciable distinction between them is the enlargement of the last joint of the abdomen in *P. Montagui*, a character which probably depends on age; the individual in question is a male, and is a little larger than the ordinary males of *P. veterum*. Milne Edwards speaks of the "female of *P. Montagui*;" being probably misled by a cursory observation of the enlarged view of the male in Leach's plate.

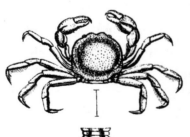

GENUS GONOPLAX, Leach.

Cancer,	Fabr. Pennant.
Ocypoda,	Bosc, Latr.
Gonoplax,	Leach, Edwards.

Generic character.—*External antennæ* long, slender, setaceous, the basal joint not notably broader than the following. *Internal antennæ* lying in transverse cells. *External pedipalps* with the third joint transversely subquadrate, the anterior inner angle truncate for the insertion of the palp. *Anterior feet* equal, extremely long in the male, nearly cylindrical; the remaining pairs somewhat compressed, the fourth pair the longest, then the third, the fifth, and the second. *Carapace* quadrate, much broader than it is long, narrowed behind; the fronto-orbitar margin extending the whole breadth. *Orbits* long, transverse, open, terminating at the external angle of the carapace. *Eyes* small, with extremely long peduncles. *Abdomen* in both sexes seven-jointed.

DECAPODA.
BRACHYURA.

GONOPLACIDÆ.

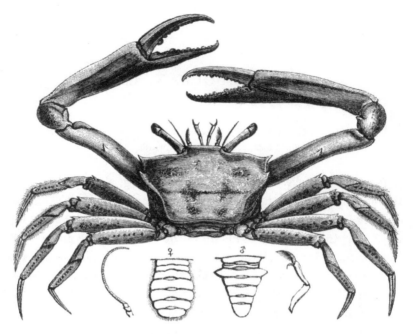

ANGULAR CRAB.

Gonoplax angulata.

Cancer angulatus,	FABR. Suppl. p. 341. PENN. Brit. VI. Zool. IV. p. 7. t. v. fig. 10. HERBST, t. i. f. 13.
Ocypoda angulata,	Bosc. Hist. Nat. des Crust. I. p. 198. LATR. Hist. Nat. des Crust. &c., VI. p. 44.
Gonoplax angulata,	LEACH, Edinb. Encycl. VII. 430. EDWARDS, Hist. Crust. II. p. 61. COUCH, Cornish Fauna, p. 72.
„ *bispinosa,*	LEACH, Malac. Brit. t. xiii.
? *Gelasimus Bellii,*	COUCH, Corn. Faun. p. 73.
? *Gonoplax rhomboides,*	ROUX, EDWARDS, &c.

THE carapace is half as broad again as it is long, broader across the anterior margin than at the posterior, rounded from before backwards, nearly even from side to side; the anterior outer angle with a prominent and acute spine, and a smaller one behind it on the lateral margin. Front entire, incurved, broad; orbits directly transverse, open directly forwards; eyes on long peduncles, and protected by the latero-anterior spines. The anterior legs in the male four times the length of the carapace; those of the female much shorter, as are those of the young male. The arm cylindrical, curved, armed with a small spine near the middle of its upper side; a still smaller spine on the inner margin of the wrist; the hand gradually increasing in size towards the extremity, rounded, somewhat flattened at the sides; fingers finely toothed, and with a few larger tubercles; in the older individuals separated for nearly half their length. The remaining feet long, slender; the second and third pairs with the last three joints hairy on the edges. Abdomen of the male triangular from the third joint to the extremity, the last joint forming nearly an equilateral triangle; of the female broadly oval: both fringed with hair.

Colour dull yellowish red. The moveable finger, in the male only, blackish.

It was not until this species was obtained by Montagu in the Estuary of Kingsbridge, Devon, that it was ascertained to be British. Since that period it has been repeatedly taken on the southern parts of the coast. I have received it through the kindness of Mr. Couch from Cornwall, and from the coast of Wales, where it was procured by Mr. Eyton; but I am not aware of its having been found on the eastern coast, nor have I heard of its having been taken in Scotland. In Ireland we have the following records of its occurrence from Mr. W. Thompson's account of the Crustacea of that portion of the kingdom. " Mr. J. V. Thompson's collection contains an Irish specimen of this Crab, marked 'rare.' Mr. R. Ball has found the species in the stomachs of cod-fish, purchased in the markets of Youghal and Dublin, and commonly in those brought to the former place: four of these Crabs is the greatest number he has obtained from the stomach of a single fish. In the Ordnance Collection is a fine example, labelled as procured at 'Bangor, January, 1836.'"

It is a Mediterranean species, and is found also on the north-west and southern coasts of France, according to the observation of Dr. Milne Edwards.

I cannot but believe that the *Gonaplax rhomboides* of Roux and other authors, is merely a variety of this species, in which opinion I concur with Mr. W. Thompson. Should further observations, however, prove that it is distinct, it is probable that the *Gelasimus Bellii* of Couch's Cornish Fauna will prove to be the female, or young male of that species.

It is found in moderately deep water; and Leach records on the authority of Cranch, that "they live in excavations formed in the hardened mud, and that their habitations, at the extremities of which they live, are open at both ends." They appear to constitute a favourite food of the cod and other fish, as, in addition to the observation of Mr. Ball quoted above, Mr. Couch states that it is often taken in their stomachs.

DECAPODA.
BRACHYURA.

GRAPSIDÆ.

GENUS PLANES, Leach.

Cancer,	Herbst, Fabr.
Grapsus,	Latr. Roux, Leach.
Planes,	Leach, Bowdich.
Nautilograpsus,	Edwards, Mac Leay, Goodsir.

Generic character.—*External antennæ* lying at the exterior of the antennary fossæ, the basal articulation nearly horizontal, extending obliquely forwards and outwards, the outer extremity the narrowest; its moveable portion very short, setaceous, the joints rounded. *Internal antennæ* folded transversely in the fossæ, which are covered by the lamellar front, and separated by a broad process extending from the epistome to the front. *External pedipalps* with the third joint broader than it is long, broadly and not deeply emarginate at the inner half of the anterior margin. *Anterior legs* robust, rounded, smooth, the hand inflated, the fingers somewhat inflected, slightly toothed; the remaining pairs much compressed. *Carapace* depressed, convex, rounded, quadrato-orbicular. *Front* broad, lamellar, bent somewhat downwards. *Orbits* distant, open above. *Abdomen* seven-jointed in both sexes; in the male acutely triangular; in the female, nearly orbicular.

This genus, the only representative of the family Grapsidæ known to have been found on our coasts, has hitherto been but very imperfectly elaborated. The synonymy of the species is much involved, and it is almost impossible satisfactorily to disentangle it. I believe there are not less than three or four species, the whole of which are found floating about amongst the sargasso or gulf-weed

Fucus vagans, or attached to the bodies of the large marine turtles. The figures of Linnæus in his "Iter Westrog."—of Bowdich in the "Excursions in Madeira and Porto Santo," the descriptions of Say, of Edwards, of Mac Leay, and others, only tend to show that there are several species in existence, but do not diminish the difficulty of distinguishing them. It is not intended on this occasion to attempt their discrimination; but it would be very desirable that the task should be undertaken by some one having the means at hand of comparing a great number of specimens. There is a good collection of them in the British Museum, and I have little doubt that I possess three species in my own collection.

I have thought it right to restore the generic name of *Planes* to these *Grapsidæ*, because it was not only applied to them by Leach in his MSS. in the British Museum, but adopted by Bowdich in his book above referred to. Whether Leach had ever published any account of the genus under the name *Planes* or not, I have not been able to ascertain; but it is highly probable that Bowdich quoted it from some such authority.

DECAPODA.
BRACHYURA.

GRAPSIDÆ.

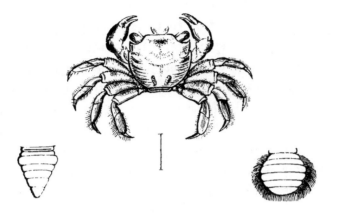

FLOATING CRAB.

Planes Linnæana, Leach.

? *Cancellus marinus minimus quadratus,*	Sloane, Nat. Hist. Faun., II. p. 270. t. ccxlv. fig. 1.
? *Grapsus testudinum,*	Roux, Crust. Mediterr. t. vi. figs. 1—6.
? *Cancer minutus,*	Fabr. Syst. Ent. XI. p. 443, ejusd. Suppl. 343. Herbst, I. t. ii. fig. 32.
? *Grapsus, ,,*	Latr. Hist. Nat. Crust. VI. p. 68.
? *,, cinereus,*	Say, Journ. Acad. Sc. Phil. p. 99.
? *Nautilograpsus minutus*	Edw. Hist. Crust. II. p. 90.
Planes Linnæana,	Leach, MSS. Brit. Mus.

The carapace in this species is nearly quadrate, with the sides somewhat rounded, and slightly contracted posteriorly: it is of a generally depressed form; the surface smooth but not polished; and there are on the posterior part of the branchial region several faint striæ, occupying the place of those which are so conspicuous in the genus *Grapsus,* and some other forms of this family. The front is lamellar, broad, projecting, slightly inclining, and entire. The orbits open above, with a small tooth at the outer angle, forming the anterior angle of the lateral margin: immediately behind this tooth is a very slight depression. The margins are very entire. The external antennæ are extremely small. The antennary fossæ are separated from the orbits only by the basal joint of the external antennæ, which scarcely fills up the hiatus. The anterior legs are robust, and, ordinarily, nearly equal; the arm is distinctly denticulate on the anterior and slightly so on the inner margin; the wrist has a minute tooth on the anterior inner and outer angles; the hand is smooth, very slightly granulated beneath, rounded and inflated; the fingers somewhat incurved, furnished with small tubercular teeth. The remaining pair of legs are considerably compressed; the upper edge of the last three joints fringed with stiff hairs; the inferior edge of the last joint, and the last but one, furnished with sharp spines, of which there are often two or three also on the upper edge of the last joint near the point, which terminates in a sharp spine. The abdomen in the male is triangular, formed of seven smooth joints, the first of which is transversely carinated; that of the female is nearly orbicular and very slightly raised along the centre.

The colour is very various in different individuals. In those which are marked in the British Museum as English, it is of an uniform brownish buff; in others grey, mottled with brown: but the most beautiful are those in which the upper parts are mottled with various shades of reddish brown and rich dark brown, with blotches of yellow or buff; the legs being marked with obscure bands of similar colours. These, however, doubtless belong to a distinct species.

The carapace in the largest specimens in my possession, which are from the gulf-weed floating in the Atlantic, is eight-tenths of an inch long, and the same broad: the females being smaller than the males. In the British specimens the length and breadth does not exceed four-tenths of an inch.

The occasional occurrence of this erratic species on our southern coast enables me for the first time to give it a distinct place in our British Fauna. There are in the British collection of Crustacea, in the British Museum, three specimens, placed there by Dr. Leach, obtained, as I believe, from the coast of Devonshire; and Mr. Couch, in his Cornish Fauna, has the following notice of another:—
" A species of the genus *Grapsus* is in the Athenæum at Plymouth, under the name of *G. pelagicus*, by Mr. Prideaux, and known to Dr. Leach. It is understood that the collection in the Museum of that Institution is confined to specimens taken on the borders of Devon and Cornwall." I have also received from this gentleman, whose diligence and tact in observing facts in Natural History is equalled by his kindness and liberality in imparting his information, a very young specimen from the Cornish coast, which is extremely small, being not more than a line in breadth. It was sent to me with some other specimens of various very small Crustacea, apparently taken from sea-weed; it is quite perfect, although so small, and is of a very pale grey colour, with small dark dots. Such is the amount of our knowledge of this species as an inhabitant of our coasts.

The several species are found in great numbers on the sargasso or gulf-weed, amongst which they breed, live, and die. One species is particularly mentioned by Sloane in his Natural History of Jamaica, as being found on the Sargasso and other submarine plants growing on the north side of that island; and adds that, "Columbus, finding it alive on the sargasso floating in the sea, concluded himself not far from some land, in the first voyage he made, on the discovery of the West Indies." They are, however, found wherever the gulf-weed floats; and it is doubtless from some accidental drifting of this plant towards our own coast, that we owe the addition of one species to the British Fauna.

As has been already observed, there are, doubtless, at least three distinct species of the genus. As the British specimens have been named by Dr. Leach, and are certainly distinct from that ordinarily found, I have thought it right to retain his name; and shall be glad to find that the investigation of the genus by some competent person has led to the adoption of sound specific characters by which the different species may be distinguished.

GENUS EBALIA, Leach.

Cancer,	Pennant, Montagu.
Leucosia,	Leach.
Ebalia,	Leach, Edwards, &c.

Generic Character.—*External antennæ* extremely minute, inserted in the inner canthus of the orbit. *Internal antennæ* lying in oblique fossæ, which are entirely separated by a small process of the epistome, and concealed by the front. *External pedipalps* elongato-triangular, reaching forwards to the margin of the *epistome;* the internal footstalk gradually acuminated, the third joint internally palpigerous. *Anterior legs* large, equal, the hand inflated, those of the male larger than those of the female; the other legs shorter than the first pair, diminishing gradually in length, terminating in a slightly curved, rather strong claw. *Abdomen* seven-jointed, but with several of the middle joints confluent; that of the male narrow, gradually diminishing from the third joint: of the female very broad, the last joint very small, abruptly narrower than the preceding. *Carapace* rhomboidal, with the angles more or less truncated or rounded; front produced, elevated. *Eyes* very small. *Orbits* with two small fissures on the superior margin.

Of this genus, which forms the English representative of the family *Leucosiadæ*, there are three distinct species found on our coasts. These are sufficiently distinct in several very tangible and essential characters; and I am surprised to find that Dr. Milne Edwards should consider them merely as varieties. The distinctions will be particularly pointed out in the descriptions of the several species. At present I am not aware that either of them has been found in any other locality than on our own coasts; but Dr. Edwards describes a species existing in the French Museum, and I have specimens from Mr. Cuming's collection from the western coast of America, which must be referred to this genus, but belonging to a new and very remarkable species. The genus was formed by Dr. Leach, who, with great propriety, separated it from his genus *Leucosia*, to which he had at first referred the species then known.

The family of which this genus forms a part is perfectly natural and well defined, and contains many very interesting forms, all of them so characteristic as to exhibit at once their close relation to each other.

DECAPODA.　　　　　　　　　　　　　　LEUCOSIADÆ.
BRACHYURA.

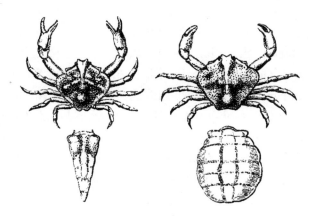

PENNANT'S EBALIA. Leach.

Ebalia Pennantii.

Specific character.—Carapace granulated, with an obtuse elevated transverse and longitudinal ridge, forming a cross; latero-anterior margin divided into two lobes by a fissure; abdomen with the third to the sixth joints united.

Cancer tuberosus,	Pennant, Brit. Zool. IV. t. ix. a, f, 19.
Ebalia Pennantii,	Leach, Malac. Brit. t. xxv. f. 1—6. Zool. Miscell. III. p. 19.

The carapace in *Ebalia Pennantii* is rhomboid, rather broader than it is long, the angles rounded, the latero-anterior margin slightly sinuous, and divided by a small fissure; the posterior margin is rounded; the front elevated and emarginate; the orbits very small, and with two small fissures above; the carapace has an elevated cross, formed by a rounded longitudinal ridge crossed by a transverse one; the whole posterior portion is elevated, and the anterior part slopes suddenly from the obliquely transverse ridge on each side; the surface is everywhere distinctly granulated. The first pair of legs are the longest and are equal; the arm is trihedrous, the wrist short and slightly inflated, the hand rounded, inflated, externally carinated, the fingers furnished with two very minute ridges on the outer surface; the whole granulated. The remaining pairs of feet are slender, the joints rounded, the terminal one slightly curved. The whole of the parts about the mouth, particularly the foot-jaws, distinctly granulated, the granulation appearing almost like minute pearls. The abdomen in the male is triangular and more than twice as long as it is broad; the third to the sixth joints united,—in the female it is much rounded, nearly as broad as it is long, the terminal articulation abruptly much smaller than the preceding, to which it is, as it were, a mere appendage.

Colour reddish brown, paler beneath, the abdomen in either sex often symmetrically spotted with red. I have a specimen obtained by Mr. McAndrew, and to whom I am indebted for it, which is all over of a lovely bright rose colour.

This species, which is the largest of the genus, is about five-eighths of an inch long, by two-thirds broad. These are the dimensions of the carapace of a female specimen in my cabinet from the coast of Devon; and Dr. Leach speaks of female specimens half as large again as his figure, which would correspond with mine, or perhaps rather exceed it. It was first described by Pennant, from specimens in the Portland Cabinet, which were probably obtained at Weymouth, a locality in which another species, *E. Bryerii*, was also first discovered. It was afterwards found on the coast of Devonshire, from whence I have obtained it, through the kindness of my friend Walter Buchanan, Esq., who procured it at Exmouth. It is mentioned in the following terms by Mr. Embleton, in his Catalogue of the Podophthalmous Crustacea of Berwickshire and North Durham. "A single specimen, taken at Redhaugh, Berwickshire, in the collection of Dr. Johnstone, and another in my own, taken in Embleton Bay, are the only ones which have fallen under my notice. In both, which are females, the abdominal covering is marked with two rows of bright scarlet spots, a character not noticed by Dr. Leach." Its occurrence as an Irish species is thus detailed by Mr. W. Thompson.* "Although this species must be considered rare, it is less so than *E. Bryerii* and *E. Cranchii*. A specimen (from Cork?) is in Mr. J. V. Thompson's collection. In September, 1836, one was dredged up from deep water in Belfast Bay, by Mr. Hyndman, and subsequently another was similarly obtained there by Dr. Drummond. Several were procured in the same locality by the collectors attached to the Ordnance Survey, who likewise dredged a specimen in Larne Loch. To Mr. G. J. Allman I am indebted for one which he found in Dublin Bay. Three examples of the *E. Pennantii* were brought up alive in the dredge from a depth of fifty fathoms, off the Mull of Galloway, by Captain Beechey, R.N."†

Its occurrence on the eastern coast of Scotland is also well attested, and I have before me an immature female specimen,‡ obtained by Mr. H. Goodsir, who notices its being generally found on stony bottoms, and on fishing-

---

* 'Ann. and Mag. Nat. Hist.' vol. x. p. 286.   † Ibid. vol. x. p. 21.

‡ The specimen here alluded to was considered by Mr. Goodsir as belonging to a distinct species; but from a careful examination of several specimens, I am satisfied that it is the present species at an immature age. The form of the abdomen is the only character in which it differs, and this has the comparatively narrowed form which always belongs to this part in the young female in all the Brachyura.

banks. Professor Forbes informs me that he has repeatedly procured it.

The above account of the localities in which this species has been found, warrants us in believing that it is not so rare as has been imagined; and that its unfrequent occurrence is to be attributed to its deep-water habits, rather than to its actual scarcity. As far as I have had opportunities of judging, females are much more numerous than males.

DECAPODA.   *LEUCOSIADÆ.*
BRACHYURA.

BRYER'S EBALIA.

Ebalia Bryerii. Leach.

Specific character. Carapace slightly and minutely granulated; lateral margin entire, somewhat revolute at the angles; two tubercles on the cardiac region, and one on each of the branchial in the male; these parts very tumid in the female. Abdomen in the male with the third to the fifth joints united; in the female, the fourth to the sixth. Arm not more than twice as long as it is broad.

Cancer tumefactus,   Mont. Trans. Linn. Soc. IX. p. 86. t. ii. fig. 3.
                          (fœm. auct.)
Ebalia Bryerii,   Leach, Mal. Podoph. Brit. t. xxv. figs. 12, 13.

THE carapace in the male is somewhat flattened, depressed in the centre, and transversely hollowed immediately behind the front, which is considerably raised, and slightly emarginate. The branchial regions and the cardiac region are raised, the elevations in the male being distinct, in the female so tumid as to form a general elevation of the whole of the posterior two-thirds of the carapace, abruptly sloping to the margin, which is turned up at the sides. The orbits are very small, and the fissures in their superior margin indistinct. The surface is minutely and almost obsoletely granulated. The arm in the male is less than twice as long as it is broad, with a projection on the inner side, and furnished on each edge with a few minute but distinct tubercles; the hand is somewhat tumid, robust, and the fingers slightly grooved. The remaining feet slender, and little different from those of the former species. The foot-jaws and other parts about the mouth, as well as the whole surface, are nearly smooth. The abdomen in the male is triangular, about twice as long as it is broad, obsoletely carinated, the third, fourth, and fifth joints united, the terminal one with a small prominent point directed backwards. In the female the general form of the abdomen much resembles that in *E. Pennantii*, but the fourth, fifth, and sixth joints are united; it is distinctly carinated.

Colour reddish white, the anterior margin and a few dots on the carapace red, with indistinct reddish bands across the abdomen in the female.

Length half an inch; breadth very little exceeding the length.

This species, which appears to be more rare than the former one, although perhaps less so than *E. Cranchii*, was first described and figured by Montagu, who at once appreciated the distinction between it and Pennant's *Cancer tuberosus*, and gives those distinctions with great discrimination. The carapace is more nearly rectangular; the whole surface nearly smooth, instead of being, as in the former case, covered with distinct pearly granulations; the three distinct tuberosities of the carapace, so different from the cruciform elevation in *E. Pennantii*, the raised margin, together with the different form and composition of the abdomen, and the more swollen and uneven character of the hands, form altogether an accumulation of distinctive characters so obvious that it is impossible to account for the two species being for a moment considered as mere varieties, as they are by Dr. Milne Edwards.

The first occurrence of this species on record is that mentioned by Montagu, who received specimens from Weymouth, where it was discovered by Mr. Bryer, to whom Dr. Leach afterwards dedicated it. This distinguished zoologist subsequently procured it through Mr. Prideaux from the Sound of Plymouth; it is mentioned by Mr. Couch in his Cornish Fauna as the only species he had himself taken. I have received both sexes from Exmouth, through the kindness of Mr. Buchanan; and I have a fine male specimen from Torquay, and a female from Tenby; for both of which I am indebted to Mr. Bowerbank, by whom they were procured by dredging. It occurs also in Mr. Bean's collection at Scarborough. Mr. W. Thompson mentions its rare occurrence as an Irish species, the only locality in which it has been found there being Belfast Bay. Captain Beechey dredged it with the former off the Mull of Galloway, in fifty fathom water.

Nothing is known of the habits of this species, nor indeed of either of the others of the genus. Its occurrence, as far as we have any data, has always been in deep water.

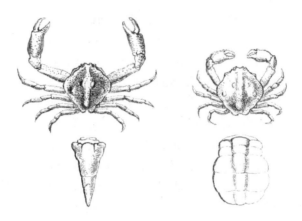

CRANCH'S EBALIA.

Ebalia Cranchii. Leach.

Specific Character.—Carapace distinctly granulated, carinated; with five tubercles, two near together on the cardiac region, two distant on the branchial regions, and one on the intestinal region; latero-anterior margin nearly entire; arm linear, three times as long as it is broad.

Ebalia Cranchii, LEACH, Zool. Misc. III. p. 20. Malac. Brit. t. xxv. f. 7—11. EDW. Hist. Nat. Crust. II. p. 129.

THE carapace in this species is more regularly rhombic than in either *Eb. Pennantii* or *Bryerii.* The surface is distinctly granulated; there is an obtuse longitudinal carina extending the whole length, and there are five distinct tubercles, of which two are very near each other on the cardiac region, one on each branchial, and a single one, larger than the others, on the intestinal. The latero-anterior margin is almost entire, having only a slight sinuation; the front is emarginate, as is also the posterior angle. The anterior pair of legs are equal, robust, and in the male nearly twice as long as the carapace; the arm is somewhat trihedrous, and three times as long as it is broad; the wrist ovate, the hand slightly tumid, the fingers shorter than the hand; the remaining pairs of legs slender, the second and third pairs in the male one-third longer than the carapace. In the female the carapace is, in proportion, a little longer than in the male, and the legs considerably shorter. The abdomen in the male has the third, fourth, and fifth joints, and the female the fourth, fifth, and sixth, united; in the former the penultimate joint is emarginate in the anterior margin to receive an angular projection in the posterior margin of the terminal joint.

Length of the carapace half an inch. Colour yellowish red, the female paler.

The male of this species so nearly resembles that of *E. Bryerii*, that without very careful examination they may very readily be mistaken for each other. The principal distinctive characters are to be found in the form and proportions of the arm, and the size of the granulations on the surface. The arm in *E. Cranchii* is three times as long as it is broad, and without any dilatation or protuberance on the inner side; in *E. Bryerii* the arm is scarcely twice as long as it is broad, and is furnished with a distinct projection on the inner side. In *E. Cranchii* the granulations which cover the surface of the body and limbs are distinct and somewhat prominent; in *E. Bryerii* they are very small, and depressed. The female in the present species very nearly resembles the male; in *E. Bryerii* the sexes are very dissimilar.

This is the most rare of the British species of Ebalia. It was discovered by the indefatigable and unfortunate Cranch, in Plymouth Sound, where it was afterwards observed, according to Dr. Leach, in considerable numbers;

it occurs in Mr. Bean's collection at Scarborough. In the Frith of Forth it is mentioned by Mr. Goodsir as being very rare. Mr. Thompson records its occurrence as an Irish species in Roundstone Bay, Connemara; Mr. Ball found several on the beach at Portmarnoch after a storm; and Captain Portlock obtained it " by deep dredging in Belfast Bay, in the course of the Ordnance Survey."*

The vignette is an illustration of the sign Cancer, from a thirteenth century drawing, contained in the Prayer-book of Queen Mary in the British Museum.

* Thompson, Ann. and Mag. Nat. Hist. vol. x. p. 285.

DECAPODA.
BRACHYURA.

CORYSTIDÆ.

GENUS ATELECYCLUS. Leach.

Cancer.	Herbst.
Cancer (Hippa).	Montagu.
Atelecyclus.	Leach, Edwards.

Generic Character.—*External antennæ* with the basal articulation very large, united to the floor of the orbit at the outer side, and to the front above, thus separating the orbit from the antennary fossa: the moveable portion inserted beneath the front, between the orbit and the antennary fossa. *Internal antennæ* lying longitudinally in the antennary fossæ, which are, as it were, excavated in the front. *External pedipalps* completely closing the buccal opening, and advancing forwards to the base of the external antennæ; the third joint much longer than broad, terminating in an oblique line, and giving attachment to the terminal portion in a notch near the middle of its internal margin. *Carapace* more or less approaching a circular form, evenly convex; the latero-anterior and lateral margins numerously toothed; the front moderately projecting, quinquedentate, the exterior tooth forming the boundary of the orbit; the hepatic regions small, the branchial very large. *Orbits*, directed forwards, with a single fissure beneath, and two above, which form a distinct tooth towards the outer angle. *Anterior legs* very large and strong, short, compressed, the hand carinated and ciliated above; the fingers curved; the remaining pairs of moderate length, compressed, the terminal joint long, acute, and nearly straight. *Abdomen* in the male, five-jointed, in the female, seven-jointed.

This genus was established by Leach for a species found by Montagu, and described by him in the eleventh volume of the Linnæan Transactions, under the name of *Cancer*

(*Hippa*) *septemdentatus*. There are now several other species known, one of which *A. cruentatus*, is found on the coast of France, and probably in the Mediterranean. It appears very nearly to resemble our species, and may possibly be a variety of it. The group is a very natural one, and its characters well defined, but its geographical distribution is so extensive as to set all ordinary laws at defiance; I have a well-marked species, hitherto undescribed, which was procured on the western coast of South America by Mr. Cuming.

DECAPODA.
BRACHYURA.

CORYSTIDÆ.

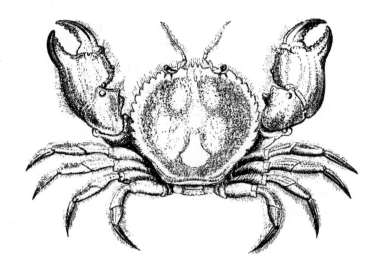

CIRCULAR CRAB.

Atelecyclus heterodon. Leach.

Specific Character.—Carapace nearly circular, the lateral margins with nine teeth, alternately larger and smaller; hairs of the legs very long.

Cancer (hippa) septemdentatus,   Montagu, Trans. Lin. Soc. XI. t. 1. f. i.
Atelecyclus   „   Leach, Edin. Encycl. VII. p. 430. Trans. Lin. Soc. XI. p. 313.
  „   heterodon   Leach, Malac. Brit. t. ii.

The general form of the carapace of this species is so nearly circular, as to distinguish it at first sight from all the other brachyurous Crabs of our coast. The lateral margins with the front form somewhat more than a semicircle, and the latero-posterior margins form three sides of a nearly regular octagon. The whole circumference is fringed with hair. The lateral margin on each side is furnished with nine teeth, which are alternately a little smaller and larger; the front is tridentate, the middle tooth being rather the longest; the whole of the teeth are slightly denticulate. The carapace is granular, moderately elevated, and the regions not very distinct. The orbits are open forwards, and have two fissures in the upper and one in the lower margin, the two former being the boundaries of a small projecting tooth. The anterior pair of legs are large and strong, compressed, and, when at rest, closing accurately against the under part of the body. The outer and upper surface of the wrist is furnished with short lines and warts of minute raised points, and there is a spine on the inner and anterior angle. The hand, which, with the fingers, is incurved, has five longitudinal lines of small raised points, besides similar ones on the superior and inferior margins. The fingers are compressed, curved, slightly toothed, and meet only at the points. The remaining legs are slightly compressed, of moderate length, and the whole are fringed with long hair. The abdomen in the male is five-jointed, nearly linear, slightly hollowed on the sides, the terminal joint triangular: in the female it is seven-jointed, very slender, being three times as long as it is broad, the terminal joint elongate and somewhat cordate.

The colour is reddish white, with red spots; the anterior feet red, the fingers black; the hair light brown.

The carapace of a full-sized male is about an inch and a quarter in diameter; the female considerably smaller.

The credit of the discovery of this species is due to Montagu, who found it on the coast of Devonshire, where it has since been found, as Leach observes, in great plenty in deep water. Mr. Couch, in his Cornish Fauna, observes that it is " common in the stomachs of fishes, chiefly codfish and rays, from the depth of twenty to fifty fathoms. They must abound at these depths, as I have found more than thirty in a single fish, and almost every ray opened for several days in succession, was found to contain them." I have obtained it from the Welch coast; and I find a very young specimen amongst some rare crustacea kindly forwarded to me from Scarborough by Mr. Bean. It has been found on the coast of Scotland, in the Frith of Forth, both by Mr. Stephenson of Edinburgh, as stated by Leach, and by my friend Mr. Harry Goodsir, who, however, states that it is rare. I have lately received a specimen which was taken from the stomach of a cod, off the coast of Zetland, by my friends Mr. M'Andrew and Professor Forbes. The accuracy and detail which characterize all the observations of my friend Mr. W. Thompson of Belfast, induce me to quote at length his account of this species as belonging to the Fauna of Ireland. " Mr. Templeton notices a Crab of this species as found by him in the stomach of a codfish, Jan. 17, 1817. In Mr. J. V. Thompson's collection is an Irish specimen, probably from Cork. In January 1839 I obtained a perfect adult male from the stomach of a brill, (*Pleuronectes rhombus,*) taken at Ardglass, County Down; it somewhat exceeds in size that figured by Leach, which again is larger than Montagu represents the species. The circumstance of the species being found in the stomachs of the cod and brill would indicate its being an inhabitant of deep water. In the Ordnance collection are examples of this Crab from Moville (Co. Donegal), Portrush, near the Giant's Causeway, and Carrickfergus. Mr. R. Ball has twice obtained it on the Dublin coast; on one occasion many specimens were found by him on the beech at Portmarnoch after a great storm." In confirmation of Montagu's and Leach's observations of the great prevalence of male specimens—those observed by the former having been all of that sex, and the latter stating that two females only were found amongst several hundreds of males, Mr. Thompson informs us that the several Irish examples which he examined with reference to their sex were all males.

The testimony which I have given from these different authors prove that the south-western coast, that of Cornwall and south Devon, is the locality in which this species is most abundant, although it occasionally occurs far to the North. That it is generally an inhabitant of deep water, is also evident; yet an observation of Mr. Thompson's would seem to show that the spawn is deposited, and that the young continue to reside, in shallower depths. "In the month of September 1835," he observes, "I obtained several small living specimens of *Atelecyclus* (carapace about two lines in length) in rock-pools, accessible at low water." Beyond these observations, we know nothing of its peculiar habits.

DECAPODA.
BRACHYURA.

CORYSTIDÆ.

GENUS CORYSTES. Leach.

CANCER.   Penn., Herbst.
ALBUNEA.   Fabr., Bosc.
CORYSTES. Latreille, Leach, Edwards.

Generic Character.—*External antennæ* very much developed, longer than the carapace, setaceous, ciliated; the basal joint thick, nearly cylindrical, inserted immediately beneath the eye in a hiatus of the orbits; the second joint also nearly cylindrical, bending downwards and inwards, approaching its fellow, so that the third joint is articulated at right angles with it, and stands forwards in contact with that of the other side—this is cylindrical, and twice as long as it is broad; the remaining joints, like the former, are nearly cylindrical, and fringed with hair. *Internal antennæ* folded longitudinally. *External pedipalps* long, narrow, standing forwards as far as the origin of the internal antennæ, leaving an aperture between themselves and the epistome directed forwards; the third joint longer than the second, and terminating forwards in a narrow and pointed process, extending beyond the origin of the fourth joint, which is articulated in a notch in the inner margin. *Anterior legs,* equal, subcompressed, with the fingers deflexed; in the male twice as long, and in the female as long as the body, the remaining legs of moderate size, compressed, ciliated; the terminal articulation very long, straight and acute. *Abdomen* in the male five-jointed, in the female seven-jointed; the first and second joints visible from above, and on nearly the same plane as the carapace. *Carapace* much longer than it is broad, elliptical, with the latero-anterior margins toothed; the rostrum triangular; orbits transverse, with two fissures above. *Eyes* little thicker than their peduncles, couching outwards and a little downwards in the orbits.

The present genus forms a very obvious approach to the division of the *Anomoura* of Edwards. Its deviations from the typical *Brachyura* are numerous and striking, and consist in the general form of the body, the relations of the buccal opening, the external pedipalps, the epistome, and the arrangement of the first joints of the abdomen, and the posterior pair of legs. There is, at present, but one species of the genus known; but it has a very nearly allied representative in the new world, the *Pseudocorystes armatus* of Edwards, discovered by M. Guy on the coast of Valparaiso, and found by Mr. Cuming and by Mr. Darwin in the same locality.

DECAPODA. CORYSTIDÆ.
BRACHYURA.

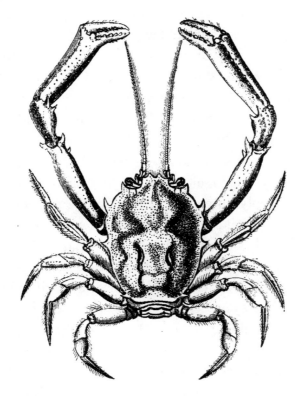

MASKED CRAB.

Corystes Cassivelaunus.

Cancer cassivelaunus,	Pennant, Brit. Zool. iv. t. vii. p. 6.
,, ,,	Herbst, I. t. xii. f. 72. Mas.
,, *personatus*	,, t. xii. f. 71. Fem.
Albunea dentata,	Fabr. Suppl. 398.
Corystes dentatus,	Latr. Hist. Nat. des Crust. et Insect. VI. p. 122. Edw. Hist. des Crust. II. p. 148. Couch, Corn. Faun. p. 74.
,, *Cassivelaunus,*	Leach, Edinb. Encycl. VII. p. 395. Mal. Brit. t. i.

The carapace in this species is longer than it is broad, in the proportion of nearly three to two; convex, with the regions somewhat distinctly marked, having a groove surrounding the cardiac and genital regions, and another short transverse depression over the intestinal region, forming altogether, in many specimens, a remarkable similitude to the features of the human face; from which circumstance I have given it the English name of "the Masked Crab." There are three acute teeth on each side of the carapace, the first forming the external angle of the orbit; the second placed on the margin of the hepatic, and the third, which is very small, on the margin of the branchial region: the surface is covered with minute scattered tufts of very short hair, scarcely distinguishable by the naked eye. The rostrum is deeply notched. The orbits are minutely granulated on the margin. The external antennæ are very long, setaceous and doubly ciliated throughout their whole length, as are also the pedipalps. The anterior feet in the male are twice as long as the body; the arm nearly cylindrical, and nearly the same length as the arm; the wrist about half as long and furnished with two spines on the inner side; the hand gradually enlarging forwards; the fingers considerably inflected, and ciliated. In the female these feet are not longer than the body; the hand scarcely longer than the wrist, and somewhat gibbous. The remaining pairs of legs are compressed, and doubly ciliated. The abdomen in the male is five-jointed; the third becoming abruptly narrower than the second; and the terminal one obtuse and rounded. In the female the first two joints are very broad; the third abruptly narrower, and, with the remaining joints, forming an oval: in both sexes this part is marginated with rather long hair.

The colour is pale red, passing into yellowish white; the arms rather deeper red. In the female the colours are much less bright and clear than in the male.

The sexes of this species differ so much from each other,

particularly in the form and development of the anterior legs, that Herbst describes them as distinct species,—an error in which he was at first followed by Latreille, who, however, afterwards corrected the mistake. It was first discovered by Pennant, who gave it the name of *Cancer Cassivelaunus*, for no very obvious reason. He gives as its habitat " the deep between Holyhead and Red-wharf, Anglesea." From the Welch coast I have also received it from Mr. Eyton; from Torquay through the kindness of Mrs. Griffiths; and it occurs in Mr. Bean's collection at Scarborough. It is generally rather a deep-sea species; and is occasionally thrown on shore " after storms or gales of wind that have been tending towards shore." In May, 1843, at Sandgate, I took a single specimen with the dredge, and on the following day ten more in the shrimp-trawl; these were all females. I have likewise obtained it at Hastings, where the late Mr. Hailstone also mentions having seen it caught by the trawlers. Mr. Couch, in his Cornish Fauna, mentions it as " scarcely common, which may be accounted for from its habit of burrowing in the sand, leaving the extremities of its antennæ alone projecting above the surface. These organs," adds Mr. Couch, " are of some use beyond their common office of feelers; perhaps, as in some other crustaceans, they assist in the process of excavation; and, when soiled by labour, I have seen the crab effect their cleaning by alternately bending the joints of their stalks, which stand conveniently angular for this purpose. Each of the long antennæ is thus drawn along the brush that fringes the internal face of the other, until both are cleared of every particle that adhered to them." As a Scottish species, it is stated by Mr. H. D. Goodsir to be rare. In Ireland it has been repeatedly taken. Mr. Wm. Thompson mentions having dredged a number of very small specimens from a sandy bottom in the open sea; and he states that the antennæ in these young individuals are much longer in proportion to the carapace than in the adult,— some, with the carapace only three lines in length, having the antennæ six lines long. The habit quoted above from Mr. Couch, of this species lying buried in the sand, with the antennæ only protruded, was also observed by Dr. Drummond, and by Mr. Ball of Dublin.

According to Mr. Hailstone's observations, the spawn is shed in April and May. I did not find any spawn attached to any of the eleven females which I took at Sandgate in the latter month.

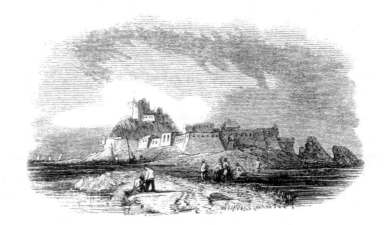

GENUS LITHODES. Latr.

Cancer,	Linn. Herbst.
Inachus.	Fabr.
Maia.	Bosc.
Lithodes.	Latr. Leach, Edwards.

Generic character.—*External antennæ* placed nearly in a line with the internal, on the outer side, and a little beneath them; the basal joint being, as it were, imbedded between the anterior margin of the carapace and a process or elongation of the lateral margin of the buccal opening, which is enclosed only at the sides, where the margins are nearly straight; the second joint is furnished with a spine on the outer side, and the third is long and cylindrical. *Internal antennæ* long, inserted beneath and somewhat external to the eyes; the first joint nearly cylindrical, thick, and bent downwards and inwards; the second and third cylindrical, slender, elongated; the terminal portion consisting of two short, setaceous, multiarticulate filaments. *External pedipalps* pediform, with the second joint short, broad, internally dilated and toothed. *Thorax* with the posterior portion free and movable. *Anterior feet* unequal, of moderate size, the fingers more or less spoon-shaped; the three following pairs very long, cylindrical; the fifth pair very small, adactylous, folded backwards beneath the latero-posterior margin of the carapace. *Carapace* cordiform, with the regions very distinct, spinous. *Rostrum* projecting horizontally. *Eyes* not enclosed within orbits, but protected externally by a strong spine; the peduncles short, approximate. *Abdomen* large, five-jointed.

This very remarkable genus was formerly placed amongst the *Oxyrhynchi* on account of the form of the carapace, which greatly resembles many of that group. Its relation to these, however, is only one of analogy; and Leach was the first to point out the discrepancies. The glimpse which he caught of its true affinities is embodied in the observation that, "in the form of its pedipalps and external antennæ, and in the position of the eyes, it approaches the Macrourous Malacostraca." It is, however, to Dr. Milne Edwards that we owe the full development of its relations, and its natural location in a group intermediate between the Brachyurous and the Macrourous forms, and in close association with *Homola*.

I have found it necessary to modify the generic characters previously given of this genus, founded as they were upon the single species hitherto described. The possession of a second, discovered by Mr. Cuming on the eastern coast of America, enables me to state that the membranous condition of the abdomen is either merely a sexual, or at most a specific distinction, as the specimen obtained by Mr. Cuming, to which I have given the name of *L. Australis*, is entirely covered with crustaceous matter.

DECAPODA.　　　　　　　　　　　　　　　　　　HOMOLADÆ.
ANOMOURA.

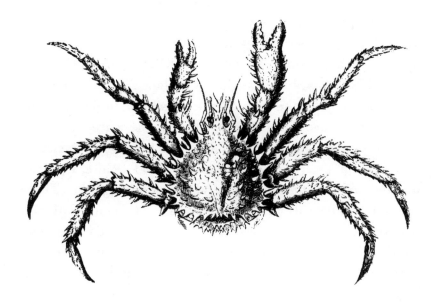

NORTHERN STONE-CRAB.

Lithodes Maia. Leach.

Specific character.—Rostrum furnished with eight spines; one above, one beneath, and one on each side at the base, two about the middle, and terminating in two, which are somewhat divergent. Carapace distinctly margined, with numerous spines longer than those on the disk. Abdomen membranaceous, with crustaceous patches representing the joints; the first and last joint entirely crustaceous.

Cancer Maia,	Linn. Syst. Nat. I. 1046, 41. Herbst. I. t. xv. f. 87, p. 219.
„ *horridus,*	Penn. Brit. IV. t. vii. f. 14, p. 7.
Inachus Maia,	Fabr. Suppl. p. 358.
Lithodes arctica,	Latr. Gen. Crust. &c. I. p. 40.　Edw. Hist. des Crust. II. p. 186.
„ *Maia,*	Leach, Trans. Lin. Soc. xi. p. 332. Malac. Brit. t. xxiv.

The form of the carapace is cordate, longer than it is broad, exclusive of the rostrum; the margin somewhat recurved all round, and beset with numerous very long and strong spines, of which those on the latero-anterior margin are regular, longer than the others, and six in number on each side, including that immediately at the outer side of the orbit. The surface is also covered with tubercles and spines; the regions distinct and elevated, excepting the hepatic, which are very slightly developed. Rostrum one-third as long as the rest of the carapace, projecting forwards, furnished with four spines at the base, of which one is placed above and another (the longest) beneath, and one on each side; two other lateral spines near the middle, and two terminal ones which are divergent. There are no distinct orbits; the eyes are contiguous at the insertion of their peduncles, and stand forwards and outwards, being protected above by the rostrum and the anterior margin of the carapace, between them by the long inferior spine of the rostrum, and at the outside by a strong spine. The anterior pair of legs are unequal, in some cases the right, in others the left being the larger; they are covered with strong sharp spines, those on the inner margin the largest; the wrist nearly cylindrical; the larger hand robust, nearly as broad as it is long; the fingers somewhat spoon-shaped, and furnished with small tufts of hair above, the opposing margins tuberculated. The second to the fourth pair of feet long, cylindrical, furnished with strong spines; the terminal joint compressed, slightly curved and acute; the fifth pair diminutive and without spines. The abdomen is coriaceous, with regular patches of crustaceous matter, representing the segments; the first is entirely crustaceous, very short, and extending quite across the breadth of the abdomen, linear and spinous; the second, fourth, and sixth joints are represented each by a pair of broad oval patches towards the margin; the third and fifth by much smaller marginal pieces between the second and fourth, and the fourth and sixth respectively; the terminal joint is also nearly oval and entirely crustaceous.

The colour is yellowish red; the spines darker, the under surface paler.

	In.	Lines.
Length of carapace	5	0
Breadth of carapace	4	0
Length of larger anterior leg	5	5
Length of leg of the third pair	8	0

This remarkable species must be considered as one of the rarer of our British Crustacea. It is, strictly speaking, a northern species, not having yet been found farther south than the Isle of Man; with the exception of a specimen in the Museum of Trinity College, Dublin, recorded to have been taken on the coast of the County Wexford. I possess, through the kindness of my friend Mr. McAndrew, several specimens, of various size, taken by him in dredging in Loch Fyne; they have also been dredged between the Isle of Man and the Mull of Galloway. The Frith of Forth (Goodwin), the Coast of Ayrshire (Thompson), of Aberdeen and of Yorkshire (Leach), are localities where this crab has at different times been obtained; and I have a specimen which was taken from the stomach of a cod on the coast of Orkney. I am uncertain at what period they cast their spawn. One of Mr. McAndrew's specimens, taken in the month of June, was carrying spawn.

The synonymy of this species has become not a little involved from some slight resemblance which it bears in its external characters to the *Maia Squinado*, and from a very obvious mistake into which Pennant has fallen in considering it identical with the *Cancer horridus* of Linnæus, the *Parthenope horrida* of subsequent naturalists. It is unnecessary to say that to the latter there is not the slightest affinity, nor even a remote external resemblance; whilst to the former the similarity is confined to the mere figure of the carapace, and the spiny armature of the body.

GENUS PAGURUS. Fabr.

Cancer,	Linn. Herbst.
Astacus,	Pennant, Degeer.
Pagurus,	Fabr. Bosc. Lam. Latr. Leach, Edwards.

Generic character.—*External antennæ* inserted in the same line with the peduncles of the eyes, and furnished with a large moveable spine, which represents the *palpus* of this organ; the last joint of the peduncle long, slender, and cylindrical; filament composed of many articulations, very long and setaceous. *Internal antennæ*, placed immediately above the ocular peduncles; the first joint nearly globular; the second and third elongate and slender, the terminal portion consisting of two *setæ*, the superior compressed, hairy; the inferior shorter, filiform. *External pedipalps* pediform, having five exserted joints; the *palpus* much developed, nearly as long as the stalk. *Anterior* feet very unequal, one of the hands being large and tumid; the second and third pairs long, ambulatory, with long curved nails; the fourth and fifth pairs small, rudimentary, sub-didactyle, the latter more distinctly so than the former. *Cephalo-thorax* membranaceous, shorter than the abdominal portion of the body. *Carapace* covering only the anterior and inferior portion of the thorax. *Abdomen* greatly developed, elongated, membranous, furnished on the upper surface with rudimentary crustaceous plates. *Tail* crustaceous, of three joints, the second joint with appendices on each side.

The family Paguridæ, and particularly the present genus, is composed of some of the most curious and anomalous forms in the whole of the class. Whilst the *Birgus*, leaving the water, and even disdaining to crawl on the ground like the true land-crab, climbs the height of the cocoa-tree, and feasts upon the young fruit, the species of the present genus clothes its soft and defenceless body in the cast-off covering of the shelled mollusks, occupying the turbinated shells of numerous species of gasteropoda, to which they closely attach themselves by means of the hooked appendages of the abdomen. From this peculiarity they have been commonly termed Hermit Crabs. Of this genus there have until lately been only two correctly distinguished species known as indigenous to this country; but the examination by Mr. William Thompson of numerous minute *paguri*, occupying various small shells found on the coast of Ireland, has led to the clear discrimination of four additional species; which, with one from the coast of Devonshire, make altogether no less than seven British species.

DECAPODA
ANOMOURA.
PAGURIDÆ.

COMMON HERMIT CRAB.

SOLDIER CRAB.

Pagurus Bernhardus.

Specific character.—Hands strongly tuberculo-granulated; terminal joints of the second and third pairs of legs spinous on the upper side, slightly tortuous.

Cancer Bernhardus,	LINN. Syst. Nat. 1049.
Astacus ,,	PENN. Brit. Zool. (ed. 8vo.) IV. t. xviii. p. 30.
Pagurus ,,	FABR. Suppl. 411. EDW. Hist. Nat. Crust. II. p. 215.
,, *streblonyx,*	LEACH. Malac. Brit. t. xxvi. f. 1—4.

The carapace in this species has the anterior margin hollowed on each side above the insertion of the eye-stalks.

forming a short obtuse-angled rostrum. The eye-stalks are short, thick, armed with a broad, flattened, oval or lanceolate tooth. The third joint of the internal antennæ scarcely extending beyond the basal portion of the external; the second joint of the latter is armed on its outer side with a sharp tooth; its palp spiniform, longer than the eye-stalks, slender, and curved. The anterior pair of legs very robust, thick, unequal, the right being ordinarily the larger,—furnished with numerous isolated tubercles, more or less spinous; the wrist, which is nearly as long as the hand, is dilated and spinous at the inner margin; the fingers obtuse and strongly tuberculated. The second and third pairs of feet spinous on the upper side, the last joint very long, strong, compressed, slightly twisted, and a little thickened towards the extremity. Posterior pairs of feet rudimentary, terminating in an extremely short, flattened pincer. Abdomen in the female furnished with four ovigerous false feet, each consisting of a basal joint, which is elongate and cylindrical, and two terminal laminar branches; the fourth much the smallest. In the male there are three false feet, composed of a basal and a double terminal joint, one finger of which is laminar and large, the other rudimentary. The terminal joint of the abdomen is notched.

The general colour is red, passing into yellow; the abdomen brown. Usual length of the adult about five inches.

I have thought it right to follow Dr. Edwards in resuming for this species its generally received name, as it is, in all probability, the one which Linnæus assigned to it, notwithstanding the doubt which led Dr. Leach to reject it, and to substitute for it the name of *Streblonyx*, in allusion to the peculiar tortuosity of the terminal joints of the ambulatory legs.

This species is extremely common, inhabiting, in the course of its growth, almost every species of turbinated shell existing on our coasts; but in its adult state requiring a habitation not smaller than the full-sized whelk, (*Buccinatum undatum*,) in which it is constantly found. Occupying, in the early stages of its growth, the small species of *Litorina*, of *Natica*, of *Buccinum*, of *Murex*, &c. When it becomes too large for its existing dwelling, it leaves it, and seeks for one not merely large enough for its present occupation, but sufficiently so to admit of a certain degree of further increase. Hence we often find individuals in shells considerably larger than would be sufficient to protect them.

It is a question of some interest whether the Hermit Crab always chooses for its habitation a shell already empty, or whether it actually kills and devours the inhabitant of one that suits its size, and then takes possession of its violated home. The latter I believe to be true, in many if not in most cases; certainly, however, not in all, as we often find the Hermit occupying an old and long-abandoned shell. But so much more generally is it found in fresh shells, that it can scarcely be doubted, even on this ground alone, that it often obtains its habitation by violence. The fishermen on the coast are fully persuaded of this; and an intelligent person of this class at Bognor assured me that the fact has often been observed by himself and others. He stated that the aggressor seizes its victim—the whelk, for instance,—immediately behind the head, and thus kills or disables it, then eats it, and finally creeps into and appropriates its vacant shell. It holds on with great force and tenacity by means of the terminal appendages; and if taken hold of when running about, which it does with great rapidity with its usurped shell attached, it draws itself in with a sudden snap, and then resists every attempt to pull it out, closing the aperture with its stout strong legs and pincers, and thus also protecting the soft membranous abdomen.

The Hermit Crabs are much employed by the fishermen (who call them " Wigs," or possibly " Whigs,") as bait for cod ; for which purpose they answer very well for immediate use, although the original possessors and builders of the house, the whelks, are much preferred for nightlines, as remaining more firmly on the hook. They are taken in great numbers in prawn-pots for this purpose.

The species is very widely distributed, and exists in every part of our coast in great numbers, being continually taken in the dredge, the keer-drag, and the prawn and lobster pots.

DECAPODA.
ANOMOURA.

PAGURIDÆ.

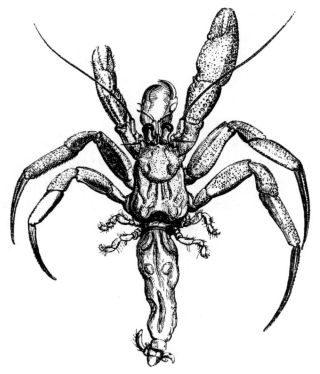

PRIDEAUX'S HERMIT CRAB.

Pagurus Prideauxii. Leach.

Specific Character.—Hands simply granulated; internal antennæ half as long again as the eye-stalks; terminal joints of the second and third pairs of legs nearly straight, grooved on each side.

Pagurus Prideaux,   Leach, Malac, Brit. t. xxvi. f. 5, 6.
   „   *Prideauxii,*   Edw. Hist. Nat. Crust. II. p. 216.

The present species resembles the foregoing in so many respects, that it had doubtless been mistaken for it by observers previous to its detection by Leach: it differs, however, from it in several well-marked characters. The anterior margin of the carapace has no median projection;

the lateral portions of the carapace are more exclusively membranous; the hands, instead of being strongly tuberculated, are merely granulated, and the wrists, on which, in *P. Bernhardus*, the tubercles become spinous on the inner margin, are in this species furnished with small tubercles; the hand and wrist are elevated along the median line of the upper surface. The ambulatory legs are nearly smooth, and the terminal joint is grooved longitudinally on each side, and is not twisted. The eye-stalks are short and very thick, and the extremity, where the eye itself is inserted, is globular. The spiniform palp of the external antennæ is more slender and less curved than in *P. Bernhardus*. It is usually of considerably smaller size, seldom exceeding two inches and a half from the front to the extremity of the abdomen.

The colour is light reddish-brown.

The discovery of this species is due to Leach, who received it from his indefatigable friend Prideaux, by whom it was taken in considerable numbers in Plymouth Sound. It has since that been found on several other parts of the coast. In Loch Fyne it has been taken by my friends Professor E. Forbes and Mr. McAndrew; and it has also been taken in Ireland by Mr. W. Thompson and Mr. Hyndman, "when dredging in Strangford and Belfast Loughs, and in the open sea off Dundrum, county Down." Mr. Thompson notices the very remarkable circumstance of its being found " in every instance inhabiting the shell invested by the *Adamsia maculata* (*Actinia m.* Adams)." And Mr. Thompson proceeds to state, " among the very numerous specimens of *Paguri* in my collection, from all quarters of the Irish coast, and found inhabiting shells of various species, not a *P. Prideauxii* occurs except in connexion with the *Actinia* already named."

In a subsequent note on the same subject, Mr. Thompson quotes Dr. Coldstream, who, in a communication in the "Edinburgh New Philosophical Journal," remarks, in treating on the *Actinia maculata* obtained by him " at Torbay, and in Rothsay and Kames Bays in Bute," that the shell which it covered was always covered by a variety of the Hermit Crab." It is also remarkable that " by Dugés the species have been found associated on the coast of France." Whether this coincidence be accidental or otherwise, it is difficult to decide without further observation; the facts, however, that Dr. Leach makes no mention of its occurrence, and that Professor Edward Forbes states that not a single specimen of the *Actinia* taken by him in the course of a season was so associated, are much in favour of their concurrence being fortuitous.

DECAPODA.
ANOMOURA.

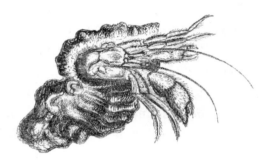

Pagurus cuanensis. Thompson.

Specific Character.—Anterior feet unequal, hisped, spinous; the palp of the external antennæ as long as the eye-stalk; their basal tooth denticulate on the inner side, half as long as the palp.

Pagurus cuanensis, THOMPSON, Report on the Fauna of Ireland, (Report of Brit. Assoc. 1843. p. 267.)

THIS is the largest of four new species of *Pagurus*, discovered by Mr. W. Thompson and named by him, though without any description or specific character, in his interesting report on the Fauna of Ireland, read before the British Association for the Advancement of Science at their session in 1843.

The carapace has the anterior margin slightly waved, without any rostral projection. The eye-stalks are very long, exceeding the basal portion of the external antennæ; the basal tooth of these half the length of the palp, straight, acute, denticulate on the inner side; the palp at least as long as the eye-stalk, curved, and furnished with long, stiff, adpressed hairs on the inner side. The anterior legs are unequal, the right the larger, covered with long stiff hair, and furnished with numerous spinous tubercles; the wrist with a row of strong short spines on the upper margin; the hand rather longer than broad, with its sides parallel and straight, furnished with rows of spines on its upper surface; the fingers short, curved, robust, strongly tuberculated. The ambulatory legs long, compressed, hairy, with the terminal joint very long, slender, and slightly curved.

Found (principally inhabiting the *Triton Erinaceus*) at Portaferry and in Bangor Bay by Mr. Thompson, and in Belfast Bay by Dr. Drummond.

DECAPODA.
ANOMOURA.

Pagurus ulidianus. Thompson.

Specific Character.—Carapace with a minute rostrum; internal antennæ the length of the basal portion of the external; anterior feet nearly equal; hand elongate, the sides parallel, roughly granulate; inner margin of the wrist toothed.

Pagurus ulidiæ,   W. Thompson, l. c.

The carapace is smooth and shining; the anterior margin hollowed over the insertion of the eye-stalks, having a very small rostriform projection in the centre. The eye-stalks thick, reaching nearly to the third basal joint of the external antennæ, with the basal tooth convex, triangular, incurved at the point. The external antennæ with the palp having a double curve, its inner margin furnished with three or four very slender long teeth. The internal antennæ about as long as the basal portion of the external. The first pair of feet somewhat unequal; the wrist roughly granular, its inner margin toothed; the hand with parallel sides, granular, gibbous, slightly carinated on the outer edge; moveable finger toothed on the outer edge. The second and third pairs of feet slightly compressed.

A very small species, so nearly resembling the young

of *P. Bernhardus* that it is difficult at first sight to distinguish them, especially as the contortion of the terminal joint of the ambulatory legs in the latter is not evident in very young individuals. The hand, however, in the present species is more elongate, its sides more nearly parallel, and the granulations on its surface more even.

It was found by Mr. Thompson at Portaferry.

DECAPODA. PAGURIDÆ.
ANOMOURA.

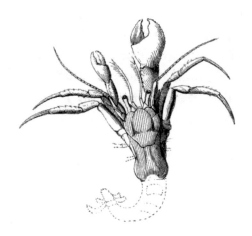

Pagurus Hyndmanni. Thompson.

Specific Character.—Anterior legs unequal; the hand oval, minutely granulated, denticulate on the outer margin; eye-stalks much shorter than the basal portion of the external antennæ; internal antennæ four times as long as the eye-stalks.

Pagurus Hyndmanni, W. THOMPSON, l. c.

THE carapace in this species is perfectly smooth, its anterior margin entire. The eye-stalks short and thick, extending only to the third basal joint of the external antennæ; the tooth at their base smooth, convex, and acutely pointed. The external antennæ about as long as the second pair of legs; the tooth on the outer side of the second joint smooth, simple, shorter than the third joint; the palp extending a little beyond the eye-stalks. The internal antennæ compressed, extremely long, being not less than four times the length of the eye-stalks. The anterior legs are very unequal, the right being the larger, granulated; the wrist of the larger with a series of small teeth along the inner margin; the hand oval, slightly convex, with an obtuse tooth at the base, the outer margin delicately and evenly denticulate; the immoveable finger broad, triangular; the moveable one carinated, minutely toothed on the outer margin, somewhat contorted in old individuals. The second and third pairs of legs slender, the joints hairy on the anterior edge; the terminal one curved.

Total length of the only specimen which could be completely examined, six-tenths of an inch; another specimen, dried in its shell, is considerably larger.

This is one of the most interesting and elegant species of this curious genus; and the perfect condition of one of Mr. Thompson's specimens enables me to give a more satisfactory description of it than of some of the others discovered by him. The form of the hand, the arrangement of the parts about the ophthalmic and antennary regions, the unparalleled proportional length of the internal antennæ, distinguish it at a glance from every other species.

It was found inhabiting *Turritella terebra* at Portaferry by Mr. Thompson, and in Belfast Bay by Mr. Drummond.

The figure is enlarged to two diameters and a half.

DECAPODA.
ANOMOURA.

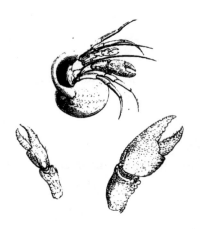

SMOOTH HERMIT CRAB.

Pagurus lævis. Thompson.

Specific Character.—Carapace with the anterior margin raised; eye-stalks short and thick, reaching to the middle of the third joint of the internal antennæ; hand minutely granulated, polished, with two obsolete teeth at the base towards the inner side, and a minute tubercle at the outer.

Pagurus lævis,   Thompson, l. c.

The carapace of this pretty species is smooth and polished, somewhat heart-shaped; the anterior margin waved, and slightly raised. The external antennæ are of moderate length, the basal tooth short, pyriform, acute; the palp doubly curved, nearly as long as the basal portion of the antennæ, slightly denticulated. The eye-stalks, short, thick, extending a little beyond the middle of the third joint of the internal antennæ, which are slender. The right anterior foot very much larger than the left; the wrist rather roughly granulated, denticulated along the inner margin, which terminates in a tooth, contiguous to which is a smooth obtuse tubercle; the hand broadly ovate, convex, slightly granulated, polished, with two small obsolete approximating tubercles near the base; the remaining second and third pairs of feet compressed; the joints carinated and slightly spinous above; the terminal joint long, slender, and slightly curved.

The general colour is yellowish testaceous, and there is a distinct red mark extending the whole length of the hand and bifurcating towards the fingers.

The specimens taken by Mr. Thompson at Portaferry were very small, and being contracted by drying and otherwise injured, afforded but little opportunity for a minute description. Having, however, recently obtained several large and perfect specimens amongst other *Paguri* from Falmouth, through the kindness of Mr. Corks, I am enabled to give the above description, and to add to the locality afforded by Mr. Thompson, that of the coast of Cornwall.

DECAPODA.
BRACHYURA.

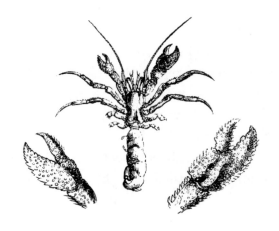

ROUGH-CLAWED HERMIT CRAB.

Pagurus Forbesii. Mihi.

Specific Character.—Eye-stalks club-shaped, as long as the basal portion of the internal antennæ; hand with irregular depressions, rough and strongly serrato-denticulate on the inner side.

OF this strongly characterised species, the carapace is subordinate, smooth. The external antennæ longer than the anterior pair of legs; the tooth at their second joint, extending to the extremity of the third joint of the basal portion; the palp nearly as long as the basal portion of the antennæ, slightly curved, and fringed with a few long hairs. The eye-stalks are club-shaped, as long as the basal portion of the internal antennæ. The anterior feet slightly unequal, the right being the longer; the wrist and hand roughly granulated; the inner margin of the wrist toothed; the hand ovate; the surface of the right with irregular depressions; the margins strongly serrato-denticulate. The second and third pairs of legs are slightly compressed, the fourth joint spinous on the upper edge; the terminal joint hairy, spinous beneath.

The whole of the legs with numerous small reddish brown spots.

This curious species differs obviously from every other inhabiting our coasts. I discovered it amongst some small *Paguri*, which I received through the kindness of Mr. Corks, of Falmouth, and which consisted of no less than four species, all obtained by him on that coast. The other species with which it was associated, were *P. Bernhardus, Prideauxii,* and *lævis.* Of the present new one there was but a single specimen.

I have named this species after my distinguished friend and colleague, Professor Edward Forbes, to whom I am indebted for several interesting additions to the objects of this work.

GENUS PORCELLANA, Lam.

Cancer.	Lin. Herbst. Pennant.
Porcellana.	Lamarck, Edwards.
Porcellana Pisidia.	Leach, Desmarest.

Generic Character.—*External antennæ* inserted at the outer side of the eyes; the basal portion formed of three joints, of which the second is the largest and the longest; the terminal portion very long, setaceous. *Internal antennæ* concealed beneath the front, very small. *External pedipalps* greatly developed; the second joint very large, rounded, with a single tooth on the outer anterior angle; the third joint much smaller, irregularly trigonal, and with the remaining joints fringed with long hair at the edges. *Anterior feet* very large, and more or less flattened; the arm very short; the wrist long and dilated on the upper and inner edge, so as to form a hollow space, in which the hand lies when retracted; the hand narrow at its base, becoming very broad forwards; the fingers strong and scarcely toothed. Second, third, and fourth pairs of *feet* ambulatory, nearly cylindrical; fifth pair very small, didactyle, and doubled together at the latero-posterior angle of the carapace. *Carapace* suborbicular, depressed. *Eyes* small, lodged in an *orbit*, the parieties of which are imperfect, excepting above. *Abdomen* very large, much developed, nearly as long as the carapace, ordinarily closed against the sternum, composed of seven distinct segments, and terminating in a broad fan-like tail, formed, as in the Macroura, of the terminal segment of the abdomen and the appendices of the penultimate.

In this genus we find a marked approach to the Macrourous group in the development of the pedipalps and of the tail, as well as in several other less obvious characters. The form of the carapace, however, recalls that of the true BRACHYURA.

After a very careful consideration, I cannot place the genus *Æglea* in a different family from *Galathea*, nor the present genus from *Æglea*; I have therefore thought it necessary to form one large family of the two families *Porcellaniens* and *Galatheides* of Edwards. They differ in few important particulars, excepting in the comparative development of the abdomen, and a corresponding difference in their natatory powers; and the general form is so similar when the abdomen of the present genus is displayed, as to show to the most casual observer how near is the affinity.

DECAPODA.
ANOMOURA.

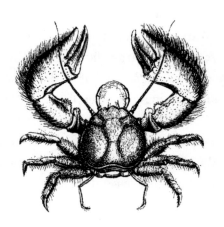

HAIRY PORCELAIN CRAB.

Porcellana platycheles.

Specific Character.—Front with three flattened triangular teeth, the middle one the longest, and slightly grooved; hands very large, hairy on the outer edge; fingers triangular; wrist with a denticulated lobe at the base.

Cancer platycheles,	PENNANT, Brit. Zool. (8vo.) IV. t. iv. f. 2, p. 9; HERBST. t. ii. f. 6.
Porcellana ,,	LAM. Anim. sans. Vert. v. p. 230.
	LEACH, Dict. des. Sc. Nat. xviii. p. 55.
	EDW. Hist. Nat. des. Crust. ii. p. 255.

THE carapace in this species is rather longer than broad, the front trifid, the middle lobe or tooth rather longer than the others, and having a slight median groove; the surface of the carapace polished, in young specimens covered with short hair, which is longer at the margin, where it is permanent; a considerable depression behind the genital region of the carapace; the orbits much arched above. The external antennæ much longer than the carapace. The anterior legs are large, flattened above ; the wrist quadrilateral, the sides nearly parallel, rather longer than broad, rounded beneath, furnished on the inner margin near the base with a triangular lobe or tooth, which is slightly denticulated ; the hand flattened above, the palmar portion triangular, furnished on the outer side with long close hairs ; the fingers triangular, slightly incurved, meeting only at the tips, the moveable one deeply grooved through its whole length, the inner edges slightly granulated. The second to the fourth pairs of feet, compressed at the sides, rounded beneath, hairy ; the terminal joint very short. Abdomen with the centre slightly raised.

Colour reddish-brown, paler and yellowish beneath ; the hairs brown.

Ordinary length of the carapace half an inch ; length of the anterior pair of legs one inch and three-tenths.

The distribution of this species is extensive, and in some localities it is also very numerous. I have received specimens from various parts of our coast, from the Orkneys to the Land's End. It is found also on several parts of the Irish coast ; and it is plentiful on the coast of France, and in the Mediterranean. Some of the largest and finest that have come under my observation, were sent me by Dr. Duguid from Kirkwall in Orkney. It is a littoral species, being generally found under stones at low water. It bites severely, as Dr. Duguid remarks ; and if seized by its claws, has the power of throwing them off instantly to facilitate its escape.

This is a further example of the favourable influence of a northern climate on the growth and development of particular animals, the specimens which are ordinarily taken on the northern part of our coast, and especially those which I received from Orkney, being much finer than those which I have obtained from the Mediterranean. A similar observation has already been made respecting *Pisa tetraodon*, and I have noticed the same difference in several other species. I have, however, lately received a very large specimen from the coast of Cornwall, through the kindness of Mr. Corks of Falmouth.

DECAPODA.
ANOMOURA

PORCELLANADÆ.

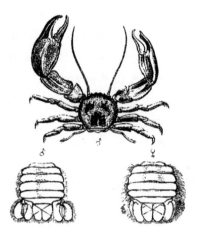

MINUTE PORCELAIN CRAB.

Porcellana longicornis.

Specific Character.—Front three-lobed, the middle lobe with a deep groove; hands unequal, long, narrow, and convex.

Cancer longicornis,	PENN. Brit. Zool. iv. HERBST, II. t. xlvii. f. 3.
Pisidia Linnæana,	LEACH, Dict. des Sc. Nat. XVIII. p. 54.
Porcellana Leachii,	GRAY, Zool. Miscell.
„ *longicornis*,	EDW. Hist. Nat. Crust. II. p. 257.

THE carapace is nearly circular, convex, nearly smooth, with a distinct thin lateral margin; the front with three lobes, the middle one so deeply grooved as to appear bifid; there are a few obsolete lateral striæ over the branchial regions. The external antennæ very long and slender; the internal of moderate length. The anterior pair of legs very unequal, the larger being, in many cases, half as long again, and nearly twice as broad, as the smaller one: in the former, the arm is very short, the wrist large and quadrate, the sides being parallel; the hand is convex, very slightly and obtusely carinated; the fingers, in the adult, touching only at the extremity, slightly tortuous; the smaller hand differs from the larger in being much more strongly carinated and grooved; the fingers are hairy on the inner edge, and the immovable one bifid at the extremity. The remaining feet are slender, and scarcely hairy at any period. The abdomen is broad, smooth, and without hair.

The colour of the carapace varies very much; it is generally pale red, frequently with irregular markings of dark reddish brown, in other specimens of bright red.

The ordinary length of the carapace is from two lines to two and a half.

This is a very pretty and a very common species. It is found under stones a little beyond low-water mark, and is very often brought up in great numbers with the oyster dredge.

I believe that the *Cancer longicornis* of Pennant, *C. hexapus* of Herbst, *Pisidia Linnæana* of Leach, *P. longicornis* of the same author, *Porcellana Leachii* of Gray, and *P. acanthocheles* of Couch, are one and the same species, varying only according to age and sex.

GENUS GALATHEA. Fabr.

Cancer, Linn. Degeer. Herbst.
Astacus, Pennant.
Galathea, Fabr. Latr. Leach, Edwards.

Generic character.—*Antennæ* inserted on the same transverse line. *External antennæ* longer than the body, the three basal segments thick, the second not longer than it is broad; the terminal filament long and slender. *Internal antennæ* inserted beneath the eye-stalks, the peduncle elongate; the last segment acute, multi-articulate, ciliated beneath. *External pedipalps* with the last two articulations neither dilated nor foliaceous. *Anterior feet* equal, or nearly so, thicker than the others, the claw well-formed; *second, third,* and *fourth* pairs of *feet* simple, alike in form, with acute nails; fifth pair spurious, very slender, doubled above the others within the branchial cavity, terminating in a rudimentary hand. *Carapace* depressed, rather longer than broad, terminating anteriorly in a sharp, more or less prominent, triangular rostrum, which covers the base of the eye-stalks. *Eyes* large and bent downwards. *Abdomen* longer and nearly as broad as the thorax, six-jointed; all of the joints without spines on the anterior margin, and terminating in a broad fan-like tail.

The species of which this genus is composed, are few in number, although it is probable that some have been confounded which are in reality distinct. The genus as it has hitherto stood, according to Edwards and other late writers, is composed of the genera *Galathea* and *Munida* of Leach; I have, however, on what appears to me to be sufficient grounds, restored Leach's genus *Munida;* which now consists of *Galathea rugosa* of Fabricius, and a new species obtained by Mr. Darwin, and through his kindness, in my possession.

The reasons which have induced me to consider, with Dr. Leach, that the forms which constitute the two families *Porcellaniens* and *Galatheides* of Edwards, are properly one group, have been already stated. Dr. Leach considered *Galathea* as the typical form of the family, as we may conclude from his giving to it the name of GALATHEADÆ. I, however, cannot but think that *Porcellana* is the typical form, and that *Galathea*, with *Munida*, *Æglea*, and *Grimotea* are aberrant, passing off towards the Macrourous type; although I cannot agree with Edwards in considering these latter as true *Macroura*.

DECAPODA. *PORCELLANADÆ.*
ANOMOURA.

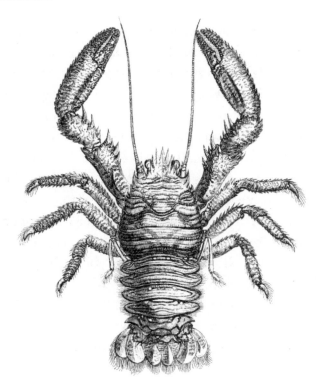

SCALY GALATHEA.

Galathea squamifera. Leach.

Specific character.—Rostrum short, with one central spine, and four on each side, the hinder one the smallest; anterior feet broad, flattened, covered with squamiform tubercles; the hands externally, and the wrists and arms internally spinous. Third joint of the external pedipalps longer than the second.

Cancer (astacus) squamifer,	Montagu.
Galathea squamifera,	Leach, Edinb. Encycl. VII. p. 393. Dict. des Sc. Nat. XVIII. p. 51. Malac. Podoph. Brit. t. xxviii. A.—Edw. Hist. Nat. Crust. II. p. 275.
	Couch, Corn. Faun. p. 77.
	Thomps. Crust. Irel. l. c. p. 105.

The carapace, exclusive of the rostrum, is a little longer

than it is broad, the lateral margins with strong acute spines directed forwards; the rostrum short, broad, triangular, terminating in a strong acute spine, and with four others on each side, of which the posterior is the smallest. The first joint of the internal antennæ short, and enlarged on the outer side, strongly spined anteriorly; external antennæ as long as the whole of the body from the rostrum to the tail. External pedipalps longer than the rostrum, the third joint longer than the second; the latter trigonous with a regular row of small teeth on the inner edge; the former with a few spines on the outer margin. Anterior feet broad, flattened, the arm and wrist strongly spined on the inner edge, without spines on the outer; the hand with smaller spines on the outer edge, none on the inner; the surface covered with small scale-like tubercles. The second, third, and fourth pairs of legs with a row of small regular spines on the anterior margin.

The general colour of this species is a greenish brown; but some which I procured at Bognor were tinged with red.

Length of the body from the rostrum to the end of the tail, three inches; such was the length of some of the specimens which I obtained on the Sussex coast, but ordinarily it is much smaller.

The first distinct account of this species appears to be that of Montagu, which Leach quotes from his MSS. It is, however, a common species all along the southern and western coast. I have specimens from Cornwall, Devonshire, Dorsetshire, and Sussex. The largest I have seen were procured by myself at Bognor, where they are often taken in considerable numbers in prawn and lobster pots. It is recorded as the most common Irish species, by Mr. Thompson, who observes that it is found on all the coasts

SCALY GALATHEA.

of Ireland. It appears to be pretty much a littoral species, occurring, according to both Dr. Leach and Mr. Couch, under stones at low tide. I have, however, taken it by the dredge in Swanage Bay, Dorsetshire, and in lobster-pots at Bognor, and Mr. Thompson mentions its being dredged by him in Ireland. This would intimate that they resort to deeper water occasionally, and Dr. Leach particularly mentions that such is the case when they are young. Those, however, which I procured in lobster-pots on the coast of Sussex were adult and remarkably large.

The vignette below is from a picture by Crome of Norwich, in the possession of Walter Buchanan, Esq.

DECAPODA.   *PORCELLANADÆ.*
ANOMOURA.

SPINOUS GALATHEA.

Galathea strigosa.   Fabr.

Specific character.—Rostrum short, with one central spine, and three on each side; anterior feet broad, very spinous on both margins; scarcely longer than the bodies; external pedipalps with the second joint longer than the third.

Cancer strigosus,   Linn. Syst. Nat. XII. 1053. Herbst, II. p. 50, t. xxvi.
Astacus   „   Penn. Brit. Zool. IV. p. 24, t. xv.
Galathea strigosa,   Fabr. Suppl. 414.   Latr. Gen. Crust. et Ins. I. p. 49.
                          Leach, Edin. Encyc. VII. p. 398.   Edw. Nat. Hist.
                          Crust. II. p. 273.
     „   *spinigera,*   Leach, Malac. Pod. Brit. xxviii.—B.

This beautiful species resembles the former, *G. squamifera*, in its general aspect, but may be distinguished from it at the first glance when adult, by the size and arrangement of the spines on the anterior feet; and on a more careful examination it may be distinguished from it at all ages, by the relative length of the second and third joints of the external pedipalps; in the present species the former, and in *G. squamifera* the latter, being the longer. The carapace is of nearly the same proportions; the rostrum has seven spines, three on each side of the central one, receding from it backwards and outwards. The lateral margin armed with strong spines. The external antennæ, with the anterior extremity of the first joint furnished with three long spines; a large spine above the auditory tubercle. External pedipalps short, scarcely extending beyond the rostrum when stretched out; the second joint much longer than the third. Anterior feet of moderate length, not much exceeding that of the whole body from the rostrum to the tail; depressed, and very spinous on all sides, excepting the outer margin of the arm. Second, third, and fourth pairs of feet, also furnished with several strong spines. Abdomen with the second and third segment unarmed. Terminal segment (the central part of the tail) much smaller at the extremity than at the base.

Colour reddish, with some blue transverse lines and spots.

Length four inches.

I have thought right to follow Dr. Milne-Edwards in considering this species as identical with Linnæus's *Cancer strigosus*, notwithstanding Leach's decided opinion to the contrary. It appears to me that the description of that species, as given by Linnæus, agrees perfectly well with our specimens of Leach's *G. spinigera*.

It is found in nearly similar localities with the former, but is certainly occasionally met with in deeper water. For the following interesting account of its habits, and of the earliest stage of its existence, as well as for the drawing of that stage of its growth from which the vignette is taken, I am indebted to my kind friend, Mr. Richard Q. Couch of Penzance, whose investigations in this and many other subjects of Natural History are well known. "This is a common species throughout the whole of the south coast of Cornwall, and I have also found it on our northern shores. It frequents pools between tide-marks, where there are loose stones and sand. It is, generally speaking, very slow in its motions, though it will frequently move with very great activity, especially when alarmed. From the great length of its first pair of legs, its motions are always retrograde. In walking its pace is tardy; but in swimming it darts from spot to spot with the rapidity of an arrow. It is never seen in any exposed part of the pool, but always seeks the shelter of stones, or some hole in the rock, so that it can retire on the least alarm. It is very remarkable to witness the accuracy with which they will dart backward, for several feet, into a hole very little larger than themselves; this I have often seen them do, and always with precision. They are laden with *ova* through the latter part of April and May, and the quantity they produce seems to be between that of the long and the short-tailed species. The *Galatheæ* are very tender, and require great care in confinement; they soon die, and hence it is not easy to rear the young. I have on many occasions hatched a very numerous family, but, like those now before me, they soon die. I can only, therefore, offer a description of them as they escape from the ovum.* As they lie in ova, the tail is bent over the thorax, and the termination rests on the space between the eyes. The tail

---

* See the figures in the next page.

is about as long as the body, slender, and composed of seven annulations; it terminates in two diverging plates. The last two annulations are of equal size and seem almost blended in one. The terminal plates are armed posteriorly with six bristles in each; the external ones are short, stout, and pointed, the others are long and slender. The carapace is rounded and indented anteriorly, and the eyes are large, sessile, and placed on a festoon of the shield. The antennæ are short, terminating in a tuft of bristles. The first pair of natatory claws are three-pointed, and, besides terminating in a tuft of setæ, are armed also along their anterior and posterior margins. The posterior claws are in three pairs, the anterior two of which are bifid; the third or posterior are small, situated at the posterior margin of the shield."

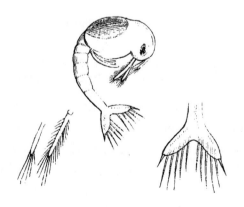

DECAPODA.
ANOMOURA.

EMBLETON'S GALATHEA.

Galathea nexa. EMBLETON.

Specific character.—Hands hairy, without spines; external pedipalps with the second joint longer than the third.

Galathea nexa, EMBLETON, in Proceedings of the Berwickshire Club.
" " THOMPSON, Annal. Nat. Hist. l. c. p. 255.

THE characters of this species approach nearly to those of the two former species, and in some respects are intermediate between them. The principal characters which distinguish it from those species, are in the armature of the hands, and the relative proportion of the different joints of the external pedipalps. Agreeing with *G. squamifera* in the absence of spines in the hand, being, in fact, more free from them than that species, and thus more especially differing from *G. strigosa,* it agrees with the latter in the comparative length of the joints of the external pedipalps or foot-jaws, of which the second joint is longer than the third; the spines of the rostrum are more flattened than in the other species. Mr. Embleton states that the "ligament of the shell" differs in colour in the three species, being bright blue in *G. strigosa*, brown in *G. nexa*, and blackish in *G. squamifera*.

There can be no doubt of the distinctness of this species. The characters above named are very constant, and the habitat is essentially different. It is also by far smaller than either of the others, as I have seen many now from various localities, none of them exceeding that figured by Mr. Embleton, and most of them much smaller. I have one specimen, with spawn, of which the thorax and abdomen together are not more than an inch in length.

It is doubtless a deep sea species. Mr. Thompson's specimens which I have before me, were obtained " from the stomachs of cod-fish brought from the coast of Down and Antrim to the Belfast market; and in Dr. Drummond's collection are specimens which were similarly procured." I have several specimens which were taken by Mr. McAndrew in dredging in Loch Fyne at a depth of from twenty to seventy fathoms, and by that gentleman and Professor Edward Forbes at Zetland.

GENUS MUNIDA. Leach.

Astacus.	Pennant.
Galathea.	Fabr. Latr. Leach, Edwards.
Munida.	Leach, Desmar.

Generic Character.—*External antennæ* with the second and third joints equal, basal joint very large; the terminal filament long and slender. *Internal antennæ* inserted beneath the eye-stalks, nearly contiguous. *Anterior* feet long, slender, and somewhat filiform, spinous within and on the upper surface. *Carapace* longer than it is broad, spinous at the margins; the rostrum, forming a long, slender, acute spine, with two smaller ones above its base, each immediately above the inner angle of the orbit, and a small spine behind each of these. *Abdomen* strongly furrowed transversely, with two or more of the segments furnished with small spines on the anterior margin; terminal joint (or central portion of the tail) nearly as broad at its extremity as at its base.

This genus was founded by Leach for the *Galathea rugosa* of Fabricius, and, as it appears, upon sufficient grounds. The general habit of the animal, the form of its rostrum, and the length and slenderness of the hands, with other characters, appeared to point out a very marked distinction between this and all other species of *Galathea* of Fabricius.

It, however, appears that no subsequent author followed Leach in considering the generic distinction satisfactory excepting Desmarest, who, however, merely followed Leach on this and other occasions. My own impressions, however, in favour of the generic separation of this form, have received an interesting confirmation, in the existence of a new and elegant species which I find among the fine collection of Crustacea procured by my friend Mr. Darwin. It possesses all the characters which I have recorded above as generic, and in all of which our own species agrees, whilst the specific distinctions are striking and obvious. I have, therefore, thought it right to restore Leach's name *Munida* to the genus, which now consists of two very distinct species.

The vignette, by Mr. C. C. Pyne, represents part of the coast between Hastings and Winchelsea.

DECAPODA.
ANOMOURA.

PORCELLANADÆ.

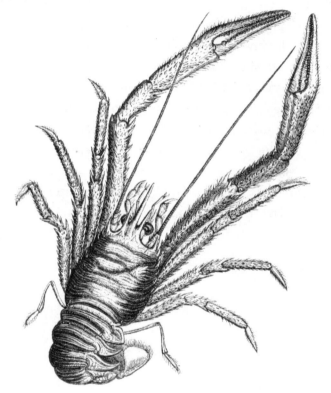

LONG-ARMED MUNIDA.

Munida Rondeletii. MIHI.

Specific character.—Anterior legs more than twice as long as the body; second and third segments of the abdomen, the former with six, the latter with four small spines on the anterior margin; the other segments without spines.

Astacus Bamffius,	PENN. Brit. Zool. iv. t. xiii.
Cancer Bamfficus,	HERBST, II. t. xxvii. f. 3.
Galathea rugosa,	FABR. Suppl. 415. 2. LATR. Hist. Nat. Crust. &c. VI. p. 198. LEACH, Malac. Pod. Brit. t. xxix. EDW. Hist. Nat. Crust. II. p. 274.
,, Bamffia	LEACH, Edinb. Encl. VII. p. 398.
Munida rugosa,	LEACH, Dict. des Sc. Nat. XVIII. p. 52.

THE carapace is slightly elliptical, transversely rugose, spinous on the lateral margins; the rostrum formed of a long styliform spine, nearly half the length of the carapace; at the base of the spine, and above it, immediately over the inner angle of the orbit, are two smaller spines, one on each side, standing directly forwards, and behind each of these another still smaller. The anterior pair of feet very long, being nearly four times the length of the cephalo-thorax, of nearly the same size throughout their length; the arms nearly three times as long as the wrist, and, as well as the wrist, armed with a series of spines on the upper and on the inner surface; the hand enlarges towards the extremity; the fingers are very slender, and longer than the hand. The remaining legs are long, slender, and cylindrical. The whole covered with close, very short hairs. The abdomen is very convex, the second segment furnished with six, and the third with four small sharp spines.

The general colour is dull reddish yellow, with redder markings.

Length of the whole animal, from the rostrum to the tail inclusive, three inches; length of the anterior legs, nearly six inches.

This remarkable species appears to be far from common on our coasts, although it is probably more numerous than has been supposed, from its frequenting deep water. It was found in Plymouth Sound by Mr. Prideaux; I have received it from Falmouth, through the kindness of Mr. Cocks; but it is not included in the Cornish Fauna by Mr. Couch. I have it also from Zetland, where it was taken by dredging by Mr. McAndrew and Professor E. Forbes; Pennant received it from Bamffshire, and hence named it *Astacus Bamffius*. Mr. Thompson, in recording its repeated occurrence on the coast of Ireland, establishes its habitat in deep water by stating several instances of its being found in the stomach of the cod; and still more remarkably, by the fact of its having been " dredged alive in water from one hundred and ten to one hundred and forty fathoms in depth, off the Mull of Galloway." These were all of them very small specimens. It is, in fact, an inhabitant of deeper water than any other of the family, not excluding *Galathea nexa*.

I have taken upon me, in restoring the genus to which this species belongs, to change the specific name. The discovery of a second species, as before mentioned, has rendered this necessary, as the latter is far more rugous in every part than the present species; and I have ventured to name it in honour of its first describer.

GENUS PALINURUS. Fabr.

Cancer, (Astacus).   Penn.
Palinurus.   Fabr. Latr. Lam. Leach, Edw.

Generic character.—*External antennæ* very thick and long; basal joint very large, and united with its fellow in front of the mouth, forming a broad epistome; the three following joints very large and spinous; the setæ but little flexible, composed of numerous short articulations. *Internal antennæ* long, the basal portion consisting of three long cylindrical articulations, terminated by two short multi-articulate setæ. *External foot-jaws* pediform. *Feet* wholly monodactyle. The first pair thicker and shorter than the others, with a spine at the termination of the penultimate joint, constituting the rudiment of a thumb. *Carapace* sub-cylindrical, with a strongly marked furrow separating the gastric from the cardiac and branchial regions; its anterior margin armed with two very strong curved spines, standing forwards over the eyes and the base of the antennæ. *Abdomen* very much developed, being thick and long: the first segment without appendages; the four following furnished with false feet, single in the male, double and hairy in the female. *Tail* very broad; the base only of each portion being crustaceous, the remaining portion membranaceous.

This genus comprises several large esculent species, one only of which is an inhabitant of our coasts. They are found principally on rocky shores, and are very widely extended. It would appear that they were well known to the ancients, and esteemed as food; and there is no doubt that the Palinurus was the καραβος of Aristotle, and the *Locusta* of Latin authors; the latter name being taken up by Belon, Rondeletius, and other writers of the earlier period of the revival of science.

This genus is the sole generic representative of the family. The characters are, however, so important and so strongly marked, as absolutely to require this distinction.

The vignette is a view on the Thames at Chelsea, by Mr. C. C. Pyne.

DECAPODA.
MACROURA.

PALINURIDÆ.

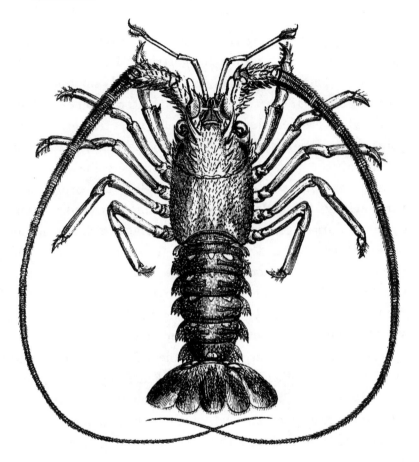

COMMON SPINY LOBSTER.

Palinurus vulgaris.   Latr.

Specific Character. — Lateral spines of the rostrum very large, triangular, smooth above, strongly dentate on the anterior margin; a strong spine at the anterior margin of the carapace beneath the orbit.

Cancer (Astacus) Homarus,	Penn. Brit. Zool. IV. t. xi. f. 22, p. 16.
Palinurus quadricornis,	Fabr. Suppl. 401.—Latr. Hist. des Crust. VI. t. lii. fig. 3, p. 193.
„ *Homarus*,	Leach, Edinb. Encyc. VII. p. 397. Trans. Lin. Soc. XI. p. 339.

Palinurus vulgaris,   Latr. Ann. du Mus. III. p. 391. Regn. Anim. Cuv. IV. p. 8. — Leach, Mal. Pod. Brit. t. xxx.—Edw. Hist.Nat. Crust. II. p. 292.

The carapace is entirely covered with spines of various size; the sulcus separating the gastric and hepatic regions from the posterior portion is deep and smooth; the central rostral tooth triangular: the lateral spines, which cover and protect the eyes, are broad, triangular, the point sharp, the anterior edge furnished with a few strong teeth; at the base of each is a strong spine: on the anterior margin of the carapace, below the orbit, is a very strong triangular spine, standing forwards. The eyes are large, globose, with a contraction immediately behind them; and the peduncle is long, exposed, and moveable. The external antennæ are extremely long; the peduncle very thick and strongly spinous; the basal joint of each meets its fellow beneath, forming a broad, smooth epistome, toothed at the anterior margin, its centre tooth much larger than the others,—at the base of this joint is placed the organ of hearing, in a tubercle raised above the surface; the remaining joints very moveable, nearly cylindrical, with strong spines on the upper and lower sides; the setaceous portion very long, composed of numerous short rings, which are closely united for the first half of its length. The internal antennæ are long, the peduncle cylindrical, the second joint nearly as long as the two succeeding ones; the setæ very short, and composed of several rings. The external pedipalps are pediform, the joints with short spines, and tufts of short stiff hairs. The first pair of feet robust in the male, smaller in the female, in both shorter than the others, monodactyle, but the penultimate joint or hand has a strong spine on the inner side, which forms a rudimentary thumb; there are strong spines on the outer margin of the other joints; the remaining pairs are strictly monodactyle and without spines; the last joint furnished with tufts of hairs. The sternum is covered with tubercles. The abdomen is nearly cylindrical, the segments smooth, terminating at the sides in a strong, flattened, triangular tooth: the first segment is without the usual appendages or false feet; those of the second, third, fourth, and fifth are simple, oval, and somewhat fleshy in the male; in the female they are double and foliaceous; those of the sixth joint form, as usual, the lateral laminæ of the tail, which are partially covered with short spines, and are crustaceous only at the outer part of the base. The central lamina or terminal segment of the abdomen is also membranaceous at the anterior part, but the whole surface is covered with numerous stronger spines.

The length of the body, from the front of the carapace to the end of the tail, is eighteen inches.

The general colour is purplish brown, with irregular dull white spots; the legs reddish white, with reddish brown irregular longitudinal bands.

This fine species is an inhabitant of our western coasts, where it occurs in great numbers, and from whence it is brought in considerable quantities to the London market. It is much esteemed as an article of food, although certainly of inferior flavour to the lobster. It is but sparingly found in the north, whether of England or Ireland, but is equally common on the southern coasts of both. It inhabits the borders of rocks, where it is often taken in crab-pots.

Its usual length is about a foot, but it sometimes reaches eighteen inches. I have taken the foregoing description from a fine male specimen of the latter size, which weighs about five pounds; and I cannot but think that Dr. Milne-Edwards is greatly mistaken in attributing to individuals of that size a weight of from twelve to fifteen pounds.

The vignette is from a sketch by Mr. Coke Smith, and represents St. Michael's Mount.

DECAPODA.
MACROURA.

THALASSINADÆ.

GENUS CALLIANASSA. Leach.

| Cancer (Astacus). | Montagu. |
| Callianassa. | Leach, Edwards. |

Generic Character.—*Antennæ* inserted in nearly the same horizontal line. *External antennæ* with an elongated peduncle, and without any moveable scale at the base. *Internal antennæ* with an elongate, rather thick, cylindrical peduncle, terminating in two setæ scarcely longer than the peduncle itself. *External pedipalps* with the second and third joints very broad, constituting when in contact a broad oval disk, and terminating in a small seta formed of the last three joints: there is no palp. *Anterior feet* very unequal, one, generally the right, being extremely large; the first three joints of moderate size; the arm furnished with a strong, hooked process on its outer margin; the wrist and hand enormously developed, the former attached to the arm by a narrow neck; these two joints are of nearly the same dimensions, and united by a straight line: second pair small, didactyle; third pair monodactyle, the penultimate joint much dilated; fourth pair simple; fifth pair subdidactyle. *Carapace* small, without any rostrum. *Abdomen* very long: the first five segments broad; the sixth abruptly narrower; the seventh triangular.

This genus may be considered as a fair type of the remarkable family to which it belongs, and which constitutes the true fossorial group of the *Macroura*. The whole of them burrow in the mud or sand, and remain generally concealed in these retreats. They are characterized by the semi-membranaceous texture of the external skeleton, by the remarkable length of the abdomen, the compressed form of the carapace, the absence of any laminar appendage to the external antennæ, and other striking characters. The family is divided into two distinct groups, according to the structure and situation of the respiratory organs. All the British species belong to that division in which the branchiæ are wholly contained within the usual branchial cavity under the margin and sides of the carapace, and which are without any branchial appendages to the under surface of the abdomen.

Of the genus *Callianassa* there is but one British species.

DECAPODA. *THALASSINADÆ.*
MACROURA.

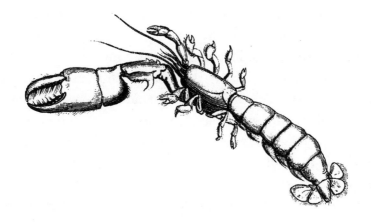

Callianassa subterranea.   Leach.

Specific Character.—Moveable finger of the larger claw, thick, obtuse; wrist and hand smooth.

Cancer (Astacus) subterraneus,   Mont. Trans. Lin. Soc. IX. t. iii. f. 1, 2, p. 89.
Callianassa subterranea,   Leach, Edinb. Encycl. VII. p. 400.—Ed. Trans. Lin. Soc. XI, p. 341.—Malac. Brit. t. xxii.—Edw. Hist. Nat. Crust. II. p. 309.

The carapace is much flattened at the sides, rounded above, very smooth; the rostrum minute. The eyes very small. The external antennæ with a long peduncle: the terminal and antecedent joints nearly cylindrical; the basal joint pyramidal. The internal antennæ have the peduncle as long as the terminal portion, which is double. The external pedipalps are rather broad, pediform; the terminal joint acute, curved. The first pair of feet very unequal: the larger, which is sometimes the left, sometimes the right, is very large, flattened, polished, ciliated on the edges; the arm is furnished, on the inner side near its origin, with a broad falciform process, the front of which is turned forwards; the wrist is quadrate, broader than it is long, connected with the arm by a narrow process, and with the hand by its entire breadth; the hand, exclusive of the fingers, is nearly equilateral, smooth, with a distinct margin at the outer side, the fingers meeting only at the point, the moveable one furnished with stiff hairs: the smaller anterior foot is very slender, the arm becoming somewhat larger at its junction with the wrist, which also enlarges towards the hand, each of these parts being longer than broad; the hand is small and smooth. The second pair of feet is didactyle, the pincers robust, and the fingers acute; the third pair has the penultimate joint transversely oval and hairy; the fourth and fifth pairs nearly filiform and simple. The abdomen is contracted at each extremity, smooth, rounded above, compressed at the sides, the second segment the longest, being as long as it is broad; the terminal or caudal segment semi-oval.

The colour of this species is a rather bright red when living, which colour it loses soon after death.

Length, about two inches.

The discovery of this remarkable species, which may be considered as the British type of the fossorial form of Crustacea, is due to Montagu who found it on the coast of Devon, where it appears to be not uncommon. It resides, as Leach states, in subterranean passages, similar to those formed by the *Gebia*. It has been found on the coast of France, and in the Mediterranean. Its claim to be considered as an Irish species is thus stated by Mr. Thompson:—" March 25th, 1839. On examining the contents of the stomach of several individuals of the *Platessa Pola,* which were taken off Newcastle (County

Down), two of the larger arms of this species, so peculiar in form, and still retaining their beautiful pink colour, were detected."*

The vignette below, by Mr. C. C. Pyne, is a view on the beach at Hastings.

* Annals, *l. c.*

GENUS GEBIA. Leach.

Cancer (Astacus).   Penn.
Gebia.   Leach.

Generic Character.—*Antennæ* inserted in nearly the same horizontal line: *external* very slender, without any vestige of a moveable scale at the base; the seta very long, its joints subelongate: *internal* very short, the double setæ rather longer than the peduncle, which is dilated on the outer side at the base. *External pedipalps* pediform, slender. *First pair of feet* somewhat robust, nearly equal, straight, the arm becoming trigonous forwards; the wrist short, rounded; the hand elongate, imperfectly cheliform; the moveable finger large, turning down to the immoveable one, which is not half its length. The remaining pairs of feet slender, slightly compressed, monodactyle. *Carapace* narrowed anteriorly, terminating in a short triangular rostrum. *Abdomen* narrowed at each extremity, somewhat depressed.

The two recorded British species of this genus so nearly resemble each other, that there is perhaps still some doubt whether they exhibit more than sexual distinctions.

DECAPODA. THALASSINADÆ.
MACROURA.

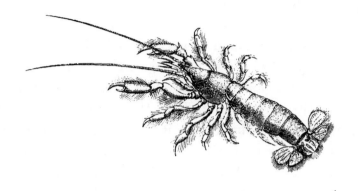

Gebia stellata.

Specific Character.—Abdomen wholly crustaceous; tail with the exterior lamella rounded, the interior subacuminate; hands with granulated hairy lines.

Cancer (Astacus) stellatus,	MONT. Trans. Lin. Soc. IX. t. iii. fig. 5, p. 89.
Gebia stellata,	LEACH, Edin. Encycl. XI. p. 400. Malac. Brit. t. xxxi. f. 1—8.—EDW. Hist. Nat. Crust. II. p. 313.

THE carapace is close at the sides, the gastric region hairy and sharply scabrous, elongate, triangular, and terminating in a small acute rostrum. External antennæ with the setæ about the length of the body. Anterior legs with the arm elongate, slightly curved, with a small tooth near the extremity; wrist very little longer than it is broad, furnished anteriorly with a sharp spine; hand three times as long as it is broad, with the moveable finger long and slender, extending far beyond the immoveable one: the second, to the fifth pair of legs, gradually more slender. Abdomen contracted at its extremities; the sides, as well as the dorsal portion, crustaceous. The tail, with the central lamina, narrowed and a little rounded forwards; the outer lamina rather longer than broad, the whole ciliated at the margin.

Length, about an inch and a half.

The discovery of this species, according to Leach, is due to Mr. Gibbs, who found it in the King's-Bridge Estuary. Montagu says that it was taken with *Callianassa subterranea*, in a sand-bank at that place; and he supposes it to inhabit the burrows formed by the *Solenes*. It is, however, not to be doubted that it forms its own burrows; and Leach states that "it has been taken on some of the shores of Plymouth Sound, under the mud of which it makes long winding horizontal passages, often of a hundred feet or more in length."

The burrowing of these fossorial species is a subject which deserves more attention than has hitherto been paid to it. The means by which it is effected are at present absolutely unknown; nor is it yet certain whether they ever avail themselves of the labours of other animals, or whether the excavations in which they are found are wholly the work of their own *hands*. The account given above, from Dr. Leach, of the extent of these passages, appear at first scarcely credible, and may well challenge a thorough examination of these points in the economy of these curious animals.

The difference of the depth which the various species of this fossorial family inhabit is very remarkable; the present species, with *Callianassa subterranea*, being found in a sand-bank, when digging for *Solenes*," whilst *Calocalis Macandreæ* was dredged from the astonishing depth of one hundred and eighty fathoms.

DECAPODA.   THALASSINADÆ.
MACROURA.

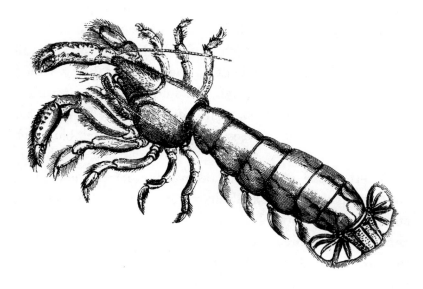

Gebia deltura.

Specific Character.—" Abdomen, with the back, submembranaceous; exterior lamella of the tail with the apex slightly rounded and dilated; the interior truncated, deltoid; the hands furnished with hairy lines."—Leach.

Gebia deltura,   Leach, Malac. Podolph. Brit. t. xxxi. fig. 9, 10.—Edw. Hist. Nat. Crust. II. p. 314.

This species, if it be indeed distinct, differs from the former, *G. stellata,* in the following particulars:—The whole animal is very much larger, sometimes not less than twice the length, and more than proportionally wider. The carapace is much broader, and more spreading at the sides. The legs are more robust; the arm of the first pair not more than twice as long as it is broad, the wrist even shorter than broad, the hand thicker, and the fingers more nearly of equal length. The setæ of the external antennæ are shorter in proportion, being, according to Leach's figure,* not more than half the length of the body. The abdomen is broader, more spread, and much less firm in its texture, the sides being almost membranaceous, and the abdominal false feet larger and more voluminous than in the other species. The different lamellæ of the tail differ also in some particulars, the exterior being rather broader than it is long, and the middle one, or terminal segment of the abdomen, nearly quadrate.† In all other respects the two species very greatly resemble each other.

Whether the distinctions above enumerated constitute anything more than sexual characters, I cannot at present determine; nor have I hitherto had access to a sufficient number of specimens to enable me to make a satisfactory comparison; but I confess I am very doubtful if it will not prove, upon further investigation, that the two British forms, and perhaps also *G. littoralis* of Risso, constitute but one species. The form and development of the abdomen, and the great volume of the abdominal false feet in *G. deltura*, are certainly very much like peculiarities belonging to the female sex, and calculated for the support and protection of the ova. " This species," says Dr. Leach, " lives with *G. stellata*, with which it was confounded until the distinctions were discovered by Mr. J. D. C. Sowerby."

* In the only specimens I have seen, which are those in the British Museum, the antennæ are somewhat injured.

† The term " deltoid " appears to be very much misplaced in describing this part.

GENUS AXIUS. Leach.

Axius.   Leach, Desmar. Latr. Edw.

Generic Character.—*External antennæ* nearly as long as the body; the peduncle furnished above with a small moveable spine. *Internal antennæ* with two setæ nearly as long as the carapace. *External pedipalps* rather slender, pediform, the joints nearly of equal length. *Anterior feet* unequal, compressed, terminated by a perfect claw; the second pair compressed, didactyle; the remaining pairs slender, compressed, simple, (the fifth pair the most slender and most compressed.) *Carapace* much compressed laterally; the rostrum triangular. *Abdomen* compressed, rounded above; the five intermediate joints of nearly equal length; the caudal joint elongate-triangular.

One species only of this genus is at present known. The genus is truly fossorial in its form, although in some respects approaching to the natatory groups of Palæmonidæ, Alphæadæ, &c.

DECAPODA.
MACROURA.

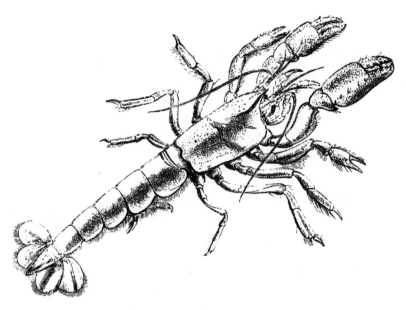

Axius Stirynchus.

Axius Stirynchus, Leach, Trans. Lin. Soc. XI. p. 343. Mal. Brit. t. xxxiii.
—Desmar, Consider. sur les Crust. t. xxxvi. fig. 1, p. 207.—Guer. Icon. Crust. t. xviii. fig. 5.—Edw. Hist. Nat. Crust. II. p. 311.

The carapace in this species is nearly semi-cylindrical, but somewhat compressed at the sides; the gastric region punctate, scabrous; the rostrum short, elongate-triangular, having a raised margin and a raised longitudinal median line. The first abdominal ring very short, furnished with a pair of rudimentary false feet; those of the four succeeding rings are fully developed, natatory, composed of a short and thick peduncle, which bears at its extremity a small styliform appendage, and two large oval laminæ, having the margins ciliated; the terminal segment is elongate-triangular; and the two pairs of lateral caudal appendages are broad, rounded, and ciliated at the margin. The first pair of legs are unequal, robust; the arm thicker anteriorly, twice as long as it is broad; the wrist broader than long, somewhat triangular; the hand thick, with nearly parallel sides; the fingers short and strong, ciliated with a few stiff hairs. The second pair of feet didactyle, rather small, compressed; the arm as long as the wrist and hand; the fingers weak; the whole, particularly the arm, furnished with long hairs at the inner margin. Of the remaining feet, which are all simple, the third pair are the thickest, and the fifth the most slender.

The following observations of the two sexes of this species are from Couch's " Cornish Fauna :"—" The male of what I judge to be the same species differs from the female in the snout, which, in my specimen of the latter, was finely notched, and without the well-marked longitudinal ridge of the former. The outer antennæ of the male are furnished with a ridge of firm hair on their inward line, decreasing towards the point, which the female is without; and the former also has well-marked brushes near the lateral edges of the abdominal rings."

Total length, three inches three lines.

General colour, pale reddish-brown.

This species, the largest of the family indigenous to this country, was first discovered by Dr. Leach " at Sidmouth, where it was taken amongst prawns on the shore. Montagu afterwards procured, near Plymouth, another specimen." I have received it from Cornwall, through the kindness of Mr. Couch, who is the only naturalist that has hitherto given any account of its habits.

" This species," says Mr. Couch, " like those of the genus *Callianassa*, has the habit of burrowing in the sand, from which it rarely emerges, and then it seeks shelter in a crevice covered with weeds, for it is sluggish in its motions, and, if distant from a soft bottom in which to sink, incapable of escaping an enemy. A female that I obtained, loaded with spawn, was dug out of the sand in the middle of summer."

It is clear that the occurrence of Dr. Leach's specimen amongst prawns must have been purely accidental, as it is essentially a fossorial species, although, as I before observed, offering some slight deviations from the typical structure of the group.

I believe it has not been found either in Scotland or Ireland. I have obtained it from the Mediterranean, and Dr. Milne-Edwards records its being indigenous to the French coast.

The subject of the vignette below is Barmouth, North Wales, by Mr. C. C. Pyne.

DECAPODA.
MACROURA.

THALASSINADÆ.

GENUS CALOCARIS. Mihi.

Generic Character.—*External antennæ* placed nearly on the same line with the internal ; the peduncle cylindrical ; the penultimate joint the longest ; a large triangular scale reaching to the end of the first joint. *Internal antennæ* with two setæ, more than half as long as the external ; the peduncle cylindrical, with the joints of nearly equal length. *External pedipalps* pediform, elongate, with a long seta. *First pair of feet* somewhat unequal, very long, compressed ; the arm slender, twice as long as the wrist, which is very short, subtriangular, flattened, with the hand somewhat gibbous, as broad as it is long ; the fingers more than three times the length of the hand, slender, much flattened ; second pair of feet didactyle, resembling the former, but much smaller ; third, fourth, and fifth pairs simple, long, and slender. *Carapace* very large, terminating in an acute triangular rostrum, from which is continued, backwards and outwards, on each side, a raised line furnished with small acute spines. *Eyes* rudimentary, subglobose, without any pigment or corneæ. *Abdomen* long, compressed, enlarged at the middle segment, contracted at each extremity ; the terminal joint, or central lamina of the tail, longer than broad, rounded.

Of this interesting fossorial form one species only is known, and it is now described for the first time. Although the structure of the feet in general, and especially of the anterior pair, together with the presence of a spinous scale on the outer side of the peduncle of the external antennæ, and some other characters, exhibit an aberration from the usual type of the fossorial family, yet its essential characters shew it to belong to that group. The most remarkable peculiarity, however, which it exhibits, is the absence of any colouring pigment and of corneæ in the eyes, to which more particular reference will be made in the specific description.

DECAPODA. THALASSINADÆ.
MACROURA.

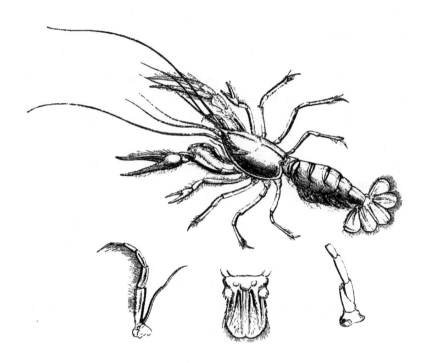

Calocaris Macandreæ. Mihi.

The crust of this species is very thin, its texture slight and flexible. The carapace is large, somewhat cylindrical, narrowed forwards, and terminating in an acute triangular rostrum, from which a raised line passes backwards and outwards, furnished on each side with four sharp flattened teeth, and inclosing a triangular space over the gastric region. A small raised medial line extends along the whole length of the carapace. The anterior feet are two-thirds the length of the whole body; the fingers very long, much compressed, and longitudinally grooved, furnished with a few small tubercles; the hand, which is gibbous, has a double carina on the upper side, which terminate each in a small spine above the orgrin of the moveable finger. The second pair of feet is distinctly and evenly didactyle, resembling in general form the first pair, but very much smaller; and the remaining pairs are very slender, and monodactyle. The whole of the feet, as well as most of the limbs and other appendages, are hairy. The abdomen is shorter in proportion, and less cylindrical, than in some other of the fossorial forms.

The general colour is a delicate pink or pale rose, varying in depth in different parts; but it soon becomes white after being placed in spirit.

The total length is about two inches.

This species constitutes one of the most singular and interesting additions which have, for a long time past, been made to our list of British Crustacea. Allied as it is in its essential characters to the *Thalassinadæ* in general, it exhibits some points of structure so abnormal, that at the first examination it would scarcely be recognized as belonging to that group. Instead of the thick and clumsy hands, the imperfect claws, and the short, solid form of the other limbs, which are exhibited in *Gebia* and *Callianassa*, we see in this species a remarkable degree of slenderness in the limbs, and an almost normal structure of the hands, assimilating it in some degree to the ordinary *Palæmonidæ* or *Astacidæ*. The absence of all colouring pigment, as well as of the corneæ in the eyes is a very remarkable, and, as I believe, an unique instance in the whole of the higher forms of Crustacea. But it is admirably in keeping with its habits, as will be presently seen.

In a fine collection of Irish Crustacea, made by my friend Mr. W. Thompson, and obligingly lent to me by him some three years since, there occurred a pair of the anterior hands of some crustacean which was wholly unknown to me, and unlike every other form I had ever seen. The only note which I found appended to them intimated that they had been taken from the stomach of a flat-fish, a ground feeder therefore, and in deep water. In the course of last year (1845) I received from my friend Mr. McAndrew, amongst some other Crustacea dredged by him in Loch Fyne and the Mull of Galloway, specimens of the present species, an examination of which at once shewed me that the claws obtained by Mr. Thompson belonged to the same animal. Mr. McAndrew and Professor Forbes have since again obtained it, and have completely established the remarkable fact, that it occasionally inhabits a depth of no less than one hundred and eighty fathoms, in which situation it is fossorial in sandy mud. Now it is clear that at such a depth, and of fossorial habits too, distinct vision would be useless and unavailing; and this at once accounts for the rudimentary character of the eyes, which are entirely white, and exhibit the appearance shewn in the vignette.

I have named it after my friend Mr. McAndrew, who first obtained it, and who has made so many important additions to our British Marine Fauna.

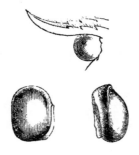

EYES OF CALOCARIS.

DECAPODA.
MACROURA.

GENUS ASTACUS. Fabr.

CANCER.	Linn.
CANCER (ASTACUS.)	Penn.
ASTACUS.	Fabr. Latr. Leach, Edw.

Generic Character.—*External antennæ* inserted beneath, and external to the internal; the peduncle thick; the second and third joints subcylindrical, covered by a moveable scale, which is broad in the middle, narrowed at each extremity, and acuminate. *Internal antennæ* with two short setæ. *External pedipalps* with the second joint very broad and thick; the terminal portion rather thick and evenly curved. *First pair of feet* unequal, tumid; the wrist short, rounded, and placed in the same line with the arm; the hands only slightly tuberculated; second and third pairs slender, filiform, didactyle; fourth and fifth pairs monodactyle. *Carapace* smooth, with a strong transverse furrow; the rostrum short, triangular, depressed, and with not more than one tooth on each side. *Thorax* with the last joint moveable. *Abdomen* very smooth; the terminal or caudal segment armed with a small tooth on each side, near the extremity, which is rounded.

A FRESHWATER genus, very properly separated by Edwards from the lobsters.

DECAPODA.   *ASTACIDÆ.*
MACROURA.

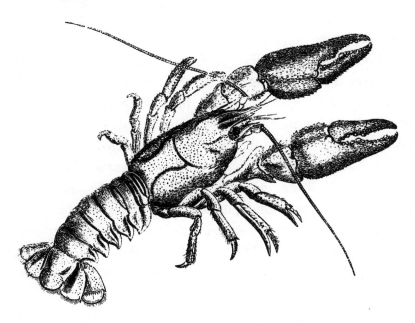

COMMON RIVER CRAYFISH.

Astacus fluviatilis. Auct.

Specific Character.—Rostrum as long as the peduncle of the external antennæ, with a slight elevation along the middle, and a small tooth on each side about one-third from the extremity. Carapace granulated.

Cancer fluviatilis,	RONDEL, Poiss. II. p. 10.
Astacus,	GESNER, ALDROV. et al. auct.
Cancer Astacus,	LINN. Syst. Nat. II. p. 1051.
Astacus fluviatilis,	FABR. Suppl. p. 406.—LATR. Hist. Nat. Crust. V. p. 235. — LEACH, Dict. des c. Nat.—EDW. Hist. des Crust. II. p. 330.

THIS, the only European species of the genus *Astacus*, may be readily known from others by the characters above given; although the general aspect of the whole of them is so similar, that they might at a cursory glance be readily confounded.

The carapace in the present species is granulated, the surface of the sides being scabrous and coarsely granulated. The rostrum is of moderate length, with a tooth on each side, about one-third from the apex: there is a slight elevation along the median line, and the margin is also distinctly raised. There is a small tooth on each side of the gastric region, near the base of the rostrum; and a spine at the anterior part of the branchial. The anterior pair of legs are thick and rounded, covered with tubercles, which become slightly spinous in some parts; the wrist denticulated on the inner margin; the hand shorter than the fingers, slightly denticulated on the inner margin; the fingers curved at the points, and strongly tuberculated on the inner edge. The second and third pairs of feet slender, didactyle; the fourth and fifth monodactyle. The abdomen is very convex, and rounded above, each segment terminating at the sides in a sharp triangular process. The terminal segment or central lamina of the tail, evenly rounded at the extremity; the lateral laminæ fan-shaped, the outer one slightly jointed about one-third from the extremity.

General length, from three to four inches.

Colour, a dull greenish gray.

Few species are more abundantly diffused than this. It is found in almost all the rivers and larger streams, not only of this country, but throughout the greater part of Europe. It is not unfrequently brought to the London market as an article of food, but is not very highly esteemed. It has long possessed a considerable degree of interest, in consequence of the facilities which it affords for watching its habits, and continuously tracing its history; and it afforded to Reaumur the means of his very interesting and original investigation into the curious subject of the moult of Crustacea; and to Rathke the subject of his observations on the growth and development of the embryo. The general facts thus observed, and their bearing upon the two interesting subjects to which they refer, will be found treated of in the Introduction to this work.

Their food consists of aquatic mollusca, the larvæ of insects, and even of small fish; and they also do not refuse any dead animals which may lie within their reach in the water. They generally appear to require the continual renewal of the respiratory fluid; and hence are generally found inhabiting running streams and rivers, in which they conceal themselves in holes in the banks. They change their crust annually, towards the end of spring, and, like all their congeners, they grow rapidly for a time after this change, and become fleshy and full.

My friend, Mr. Ball of Dublin, has favoured me with the following amusing and graphic account of an individual of this species, which he kept in confinement:—" I once had a domesticated cray-fish, *Astacus fluviatilis*, which I kept in a glass pan, in water not more than an inch and a half deep; previous experiments having shewn that in deeper water, probably for want of sufficient aëration, this animal would not live long. By degrees my prisoner became very bold; and when I held my fingers at the edge of the vessel, he assailed them with promptness and energy. About a year after I had him, I perceived, as I thought, a second cray-fish with him; on examination, I found it to be his old coat, which he had left in a most perfect state. My friend had now lost his heroism, and fluttered about in the greatest agitation. He was quite soft; and every time I entered the room, during the next two days, he exhibited the wildest terror. On the third he appeared to gain confidence, and ventured to use his nippers, though with some timidity; and he was not yet quite so hard as he had been. In about a week, however, he became bolder than ever; his weapons were sharper, and he appeared stronger, and a nip from him was no joke! He lived in all about two years, during which time his food was a very few worms, at very uncertain times; perhaps he did not get fifty altogether. I presume some person, presuming to poach in his pond, was pinched by him, and plucked him forth, and so falling, he came by his death."

Mr. Ball adds elsewhere, "The water was never changed, but some was occasionally added to supply the loss by evaporation." The truth is, that many Crustacea will live in the atmosphere, as long as they have access to water in which to bathe their branchiæ, and thus preserve them in a moist and respirable condition; but die from asphyxia when confined beneath a small quantity of water, from which the air is soon exhausted.

GENUS HOMARUS. Edw.

CANCER.	Linn.
CANCER (Astacus.)	Penn.
ASTACUS.	Fabr. Latr. Leach.
HOMARUS.	Edw.

Generic Character.—*External antennæ* placed above and to the outer side of the internal; the laminar appendage dentiform, dilated on the inner side, scarcely covering the penultimate joint of the peduncle. *Internal antennæ* with the peduncle nearly as long as that of the external. *External pedipalps* pediform, extending forwards beyond the peduncles of the antennæ. *Anterior legs* very robust, unequal; the larger with the fingers strongly tuberculated on the grasping edge, the smaller merely toothed; second and third pairs didactyle; fourth and fifth pairs monodactyle. *Carapace* nearly cylindrical, the rostrum armed on each side with three or four teeth. *Thorax* with the last joint immoveably connected with the preceding one. *Abdomen* nearly cylindrical, the segments terminating laterally in a large flat triangular process; terminal segment (or central lamina of the tail) armed with a tooth on each side near the extremity. *Tail*, with the exterior lamina divided transversely, about one-third from the extremity, with a distinct moveable joint.

DECAPODA.
MACROURA.

ASTACIDÆ.

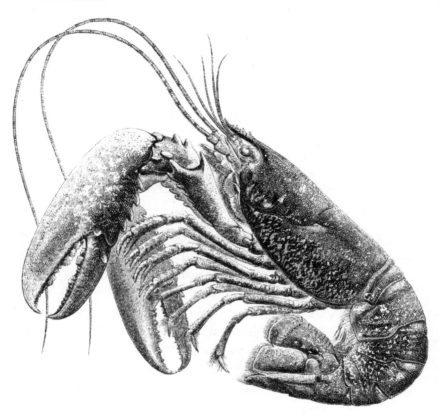

LOBSTER.

Homarus vulgaris. Edw.

Specific Character.—Rostrum extending beyond the peduncle of the external antennæ, armed with two or three strong teeth on each side, without teeth on its under surface.

Cancer gammarus,	Linn. Faun. Suec. 2033. Syst. Nat. Herbst. II, p. 42, t. xxv.
Astacus marinus,	Fabr. Suppl. 406.—Penn. Brit. Zool. IV. t. x. f. 21.
,,	Latr. Hist. Nat. des Crust. VI. p. 233.
Homarus vulgaris,	Edw. Hist. des Crust. II. p. 334.—Couch, Corn. Faun.

The body is thick and rounded; the cephalo-thorax deeper than it is broad, somewhat compressed at the sides; the surface slightly punctated: a furrow separates the gastric from the posterior regions. The rostrum projects forwards as far as the peduncle of the external antennæ; it terminates in a strong point, and has about four teeth on each side, diminishing in size backwards. There is a small tooth on each side, just behind the base of the rostrum. External antennæ with the peduncle nearly cylindrical; its base armed with a strong tooth. Eyes globular, smaller than the peduncle. Abdomen semi-cylindrical. The segments smooth, terminating on each side in a strong flattened triangular plate. The tail broad; the external lamina strongly divided at its anterior third; the margin of its posterior portion closely dentated: two strong teeth at the common peduncle of the two outer laminæ. Anterior legs very large, unequal, the larger one furnished with very strong tubercles on the prehensile edge of the fingers, which is irregular; the smaller one with the edge of the fingers straight, and having numerous small teeth; the hands with the inner margin furnished with strong white teeth; and the wrist with a few similar ones. The remaining legs filiform and weak; the second and third pairs didactyle, the fourth and fifth monodactyle.

General colour dull pale reddish-yellow, spotted with bluish-black; the spots coalescent on the upper parts.

The esteem in which this species is universally held as a delicate article of food, and the multitudes which are annually taken and brought to our markets, render it perhaps the most interesting and important of the whole class, in a commercial point of view. Lobsters are taken on various parts of our coast, particularly on rocky shores. From the southern and western coast of England a considerable number are constantly sent off to the London markets, by the South-Western Railway from Southampton, and by the Great Western from Bristol; also by steamers from Guernsey and Jersey; and again from the coast of Ireland to Liverpool. From the coast of Scotland and the Orkney and Lewes Islands, it is computed that not less than 150,000 reach the market at Billingsgate; but the principal supply is from Norway, from whence we derive not less than 600,000. There is often in the season a supply at Billingsgate of not less than from 20,000 to 25,000 lobsters in one day.* If we allow only as many to be eaten in the whole of England besides as in London, the multitude which are consumed in the course of every year is enormous.

The period in which this immense sacrifice to crustacean gastronomy principally takes place is from March to August; but it is a mistake to suppose that the lobster is only in season during that time. During the latter part of August and the following month, the lobsters are shedding their coat, and the new covering is becoming indurated; but after that time they feed ravenously, and soon become plump and firm; so that in the winter they are probably in as high flavour, and as solid for food, as during the period when they are most in request. Mr. Saunders informs me that he has reason to suppose the lobster to be very stationary, seldom wandering fifty miles from the spot of their birth; and he adds, what one would scarcely have supposed probable, that " they are as varied in appearance and character as a white man and an African." "I could tell by looking at them," says Mr. Saunders, " the part from whence they are brought." This curious fact is corroborated by Mr. Couch, who, in his " Cornish Fauna,"

* For these details I am principally indebted to Mr. J. E. Saunders, the respectable fish-salesman of Thames Street.

has the following observation:—" Lobsters do not stray far from their haunts, and hence the discovery of a new station is a fortunate circumstance for the fisherman; and each situation is found to impress its own shade of colour upon the shell."

Lobsters are frequently caught in pots, similar to those which are employed for the capture of crabs, and by somewhat similar means; but in some localities the pots are differently shaped, being formed of nets, which are held in a nearly cylindrical form by three hoops, one at each end, and one in the middle. At one end the trap is closed; at the other it is entered by a funnel-shaped prolongation of the net inwards, like some rat-traps. Mr. Thompson informs me of the capture of a lobster by means of a hook and line, baited with a whelk, which was used for taking cod. Whether the hook was taken I am not informed; but it is perhaps most probable that the lobster held firmly to the bait itself, and suffered itself to be dragged out of the water rather than quit its hold.

It is a well-authenticated and indubitable fact, that the lobster, as well as the common crab (*Cancer pagurus*), and several other species of Crustaceans, not only shed their claws and other limbs in case of severe injury to them, but voluntarily. On being seized by one of their limbs, the captive member is left in the possession of the captor, and the animal escapes, leaving his arms on the field of battle; and it is also well known to fishermen and other practical persons, that the same loss of limbs will take place in violent thunder-storms. In the words of the intelligent correspondent to whom I have already had occasion to express my obligations, Mr. Saunders, " they shoot their claws, especially after a thunder-storm or the report of cannon, and whole voyages are destroyed by this means. If time were given new claws would be formed. It is a voluntary act, and does not injuriously affect the animal." The following remarks on this subject, by my observant and accurate friend Mr. Couch, will be read with much interest, and I need offer no apology for their extent.

Mr. Couch first speaks of the effects of injuries to the antennæ, and observes that it is an erroneous opinion that these organs are ordinarily thrown off in consequence of violence done to them, and afterwards renewed. "I have not," he proceeds, "found this to be the fact; but, subjecting the parts to blows or fracture, both in short and long-tailed Crustaceans, I have found the creature suffering acutely from the injury, most so when just emerged from the water; but in no case have they rejected the whole organ in consequence of the violence. If, however, it be violently handled, a separation takes place at the terminal joint of the peduncles, in preference to any other place; and from this wound no stream of blood flows, but a fine membrane quickly forms on the surface, by which all effusion is prevented. This preservative process resembles that which takes place in case of the loss of the legs, and for the same purpose; for crabs and lobsters soon bleed to death, if the hæmorrhage be not restrained. It is only the legs, including those bearing the *chelæ* or nippers, that are readily and willingly thrown off by the animal; and in some cases, as in *Porcellana platycheles*, this is not only done on the infliction of violence, but as if to occupy the attention of some dreaded object, while the timid creature escapes to a place of safety. The general method of defence is to seize the object with the pincers, and while these are left attached, inflicting, by their spasmodic twitchings, all the pain they are able to give, the crab, lightened of so great an incumbrance, has sought shelter in its hiding-place. It is by the short and quickened muscular action of the limb itself, and not by any effort of the body or peduncle that this is effected; as the convulsion will continue for a considerable time after the separation, it follows that the twisting off of the claw, where the animal has seized human flesh for instance, or any other sensible object, is the direct way to increase the violence of the grasp. Any or all the legs may be thrown off on the receipt of injury, but not with equal facility in all the species; for in some, as in the common crab, if they be crushed or broken without great violence, they are sometimes retained, and the creature will in no long time bleed to death. To save the crab the fishermen proceed to twist off the limb at the proper joint, or give it a smart blow, when it is rejected; and in either case the bleeding is stopped. Fracture of the crust at the extreme points of the legs is not much regarded; for, being filled with an insensible cellular membrane, no violent action is excited in the muscular structure, and the part seems capable of some attempt at restoration, at least sufficient to render the evil endurable until the period of a general renewal of the surface.

After the loss of a limb, a considerable time elapses before any attempt at restoration is visible: but under some circumstances the process is much accelerated; and while it is advancing, it is commonly found that the flesh of the creature is unusually flaccid and watery. In the most common species, the first appearance of the new limb is in the middle of the scar, from whence proceeds a soft member of minute size, doubled on itself, but with all the proper proportions, and enclosed in an exceedingly fine membrane, by which it is bound down. Much of the first stage of the growth of the new limb is accomplished before it acquires density; but when the crust is rendered firm, the nutrition no longer proceeds through the encasing membrane; which a slight motion of the limb lacerates, and the leg extends to its natural position; but it continues for a long time of a much smaller size than the corresponding one of full growth, sometimes also appearing as if distorted, either from deficient nourishment, or from injury received in its unprotected state."

I have omitted from this interesting detail some speculations of the observant author, and some statements respecting which he himself speaks doubtfully; and it appears to me that it contains by far the most satisfactory and most simple statements of this interesting fact that have ever appeared. Although Mr. Couch's observations were chiefly made upon brachyurous forms, there is no doubt that the process is precisely similar in all the higher forms of Crustacea.

The reproduction of the lobster would be multitudinous, were not the young destined to become, in myriads, the prey of fish of various descriptions.

The metamorphosis of this species has been examined with care by several naturalists, and particularly by my friend Mr. R. Q. Couch. The details, as far as they belong to the general subject, will be found in the Introduction.

If the following statement, with which I have been favoured by Mr. Peach, be correct, it proves that the attachment of these creatures for their progeny does not cease on the deposit of their spawn, but continues, in a very pleasing and interesting manner, much longer than in many animals of a higher grade of organization.

"I have heard the fishermen of Goran Haven say that

they have seen in the summer, *frequently*, the old Lobsters with their young ones around them; some of the young have been noticed as *six inches long*. One man noticed the old Lobster with her head peeping from under a rock, the young ones playing around her: she appeared to rattle her claws on the approach of the fisherman, and herself and young took shelter under the rock; this rattling, no doubt, was to give the alarm. I have heard this from several, some very old men, who all speak to this *without concert*, and as a matter of course; and they are men I *can readily believe.*"

GENUS NEPHROPS. Leach.

Cancer.	Linn. Herbst.
Astacus.	Fabr. Penn. Latr.
Nephrops.	Leach, Edwards.

Generic Character.—*External antennæ* with the scale of the first joint extending to the extremity of the peduncle. *Internal antennæ* with the basal joint very broad, triangular, terminating in two *setæ*, the superior much thicker than the inferior, and slightly compressed. *External pedipalps* much elongated, the second joint the longest. *Eyes* very large, reniform. *First pair of feet* very long, unequal; the hands quadrangular, the angles carinated and strongly toothed; the fingers armed with strong tubercles: these and the second and third pairs are didactyle; the fourth and fifth monodactyle. *Abdomen* semi-cylindrical, sculptured; the lateral processes of the segments laminar and thin. *Carapace* terminating in a long rostrum, strongly dentate on each side.

This genus consists but of a single known recent species. It is nearly allied to *Homarus* and to *Astacus*, and may be considered as in some respects intermediate between them.

DECAPODA.　　　　　　　　　　　　　　　ASTACIDÆ.
MACROURA.

NORWAY LOBSTER.

Nephrops norvegicus. Leach.

Cancer norvegicus,	LINN, Syst. Nat. I. 1058.—HERBST, II. t. xxvi. f. 3.
Astacus ,,	FABR. Ent. 418.—PENN. Brit. Zool. (8vo.) IV. t. xiii. f. 1, p. 23.
Nephrops ,,	LEACH, Edinb. Encycl. VII. p. 400.—ID. Trans. Lin. Soc. XI. p. 344.—Malac. Pod. Brit. t. xxvi.—EDW. Hist. Nat. Crust. II. p. 336.

THE body of this elegant species is elongated and sub-cylindrical; the cephalo-thorax compressed at the sides; the surface slightly pubescent: the gastric region is armed with seven lines of points, of which the outermost are not more than three or four in number; the inner pair converge towards the rostrum and pass into a double carina which extends to its extremity. The rostrum extends beyond the peduncle of the external antennæ, and is armed on each side with three oblong teeth; it is ciliated on each side beneath. The posterior portion of the thorax has three lines of small points: a strongly marked sulcus runs within the posterior margin. The eyes are remarkably large and reniform; the peduncles very small at their origin, becoming suddenly much larger. The peduncle of the external antennæ is nearly as long as the rostrum: the first joint has a triangular spine at the outer side; from the anterior margin of this joint arises the broad falciform scale, which extends forwards to the extremity of the peduncle. The basal joints of the internal antennæ are very broad and laminar. The first pair of feet are very long, unequal, in some cases the right, in others the left being the larger: the arm is slender, enlarging towards its anterior extremity, carinated above and below, and armed with a few teeth: the wrist, which is short, is armed above with strong teeth, and is strongly carinated: the hand is distinctly four-sided, strongly carinated; the carinæ armed with tubercular teeth, the upper in a single, and the others in a double series; the intermediate spaces concave, and slightly pubescent: the fingers are armed with strong tubercles, particularly those of the larger claw, and the moveable one is toothed on its outer margin. The other legs are filiform, slender, and smooth; the second and third pairs being didactyle, the fourth and fifth monodactyle. The abdomen is long, each segment being beautifully sculptured; the raised portions smooth and polished, the depressions covered with a short but dense pubescence. The epimeral portion of the first abdominal segment is small and rudimentary; the second is very broad and subquadrate; the remainder are acutely triangular. The tail is very broad, and the outer lamina is slightly divided transversely at its anterior third.

The general colour is pale flesh, rather darker in parts; the pubescence light brown.

The length of the body from the tail to the rostrum is from seven to eight inches.

This is certainly one of the most beautiful of the larger *Macroura*. It is to be considered generally as a northern species, but I have received fine specimens from the Mediterranean. It is found on the coast of Norway in considerable quantities; it is also taken on the coast of Scotland, and is not unfrequently sold in the Edinburgh and other northern markets. I have occasionally seen it at the shops of London fishmongers. It is said to be a very delicate and well-flavoured food.

Although, as I have mentioned above, I have obtained it from the Mediterranean, (Dr. Milne-Edwards also records it as being taken in the Adriatic,) yet its general range is certainly confined to northern limits. Mr. Embleton says that it is not uncommon on the coast of Berwickshire, but is rarely seen farther south. Leach names only the Frith of Forth as its habitation. Mr. M°Andrew procured it by dredging in Loch Fyne. On the Irish coast it has been taken in Belfast Lough, according to Mr. Templeton. Mr. Thompson says, "I have heard of its being taken near Portaferry, about the entrance to Strangford Lough, and that it has been procured in numbers off Dundrum on the Down coast." He adds, "It is brought in great quantities to Dublin as an article of food;" and in a letter recently received from the same gentleman, he mentions its having been found in the stomach of the cod, "near Donaghadee, county Down, and also at Dungarvon, county Waterford." According to Mr. R. Ball, it is very numerous in Dublin Bay; and he has taken it from the stomachs of cod bought at that place.*

It is not included in the "Cornish Fauna" by Mr. Couch, nor have I ever heard of its appearance on that part of the coast. Mr. Thompson states that he has received specimens from Holyhead by Captain Fayrer, R.N.

* Thompson on the Crustacea of Ireland, *l. c.*, p. 209.

DECAPODA.
MACROURA.

CRANGONIDÆ.

GENUS CRANGON.

ASTACUS.   Herbst, Penn.
CRANGON.   Fabr. Latr. Leach, Edw.

Generic Character.—*External antennæ* situated nearly on the same line with the internal, on the outer side, and a very little beneath them. *Internal antennæ* dilated at the base, and having at the outer side a broad scale; the peduncle short, and terminating in two filaments. *External pedipalps* pediform; the terminal joint obtuse and flattened. *First pair of feet* subdidactyle, stronger and thicker than the others; the hand flattened, the moveable finger inflexed upon the hand, and meeting a rudimentary thumb: the second and third pairs very slender; the second didactyle; the fourth and fifth shorter and thicker than the former. *Carapace* depressed, and with only a rudiment of a rostrum. *Abdomen* large and rounded. *Branchiæ* only seven on each side.

THE family to which this species belongs is distinguished from all others by the insertion of the two pairs of antennæ on the same line, and the subcheliform structure of the anterior hand. I am inclined to follow Dr. Milne-Edwards in restoring to this genus Leach's genus *Ponotphilus* and Risso's *Egeon*, separated unnecessarily from *Crangon*, which, in fact, constitutes the only known genus of the family. It may, however, be conveniently divided into two sections, in one of which the second pair of feet is as long as the first, in the other it is not much more than half as long. The first constitutes the genus *Crangon* of Leach; the second *Poutophilus* of the same author.

DECAPODA.
MACROURA.

CRANGONIDÆ.

COMMON SHRIMP.

Crangon vulgaris. Fabr.

Specific Character.—Carapace and abdomen smooth, excepting a small spine on the median line of the gastric region, and one on each branchia; second pair of feet nearly as long as the third.

Astacus Crangon,	HERBST, II. p. 57, t. xxix. fig. 3, 4.—PENN. Brit. Zool. IV. t. xv. f. 30.
Crangon vulgarin,	FABR. Suppl. 410.—LATR. Hist. Nat. des Crust. VI. p. 267, t. lv. f. 1, 2.—LEACH, Malac. Brit. t. xxxvii. B.—EDW. Hist. Nat. Crust. II. p. 341.

THE carapace in this species is large, rounded, somewhat depressed, particularly towards its anterior part: there is no rostrum, but a slight elevation on the median line, between the eyes; a minute spine directed forwards over the gastric region, and one a little more conspicuous on each branchia. The eyes are conspicuous, naked, and not very distant. The external antennæ have the peduncle about half the length of the moveable plate, which terminates in a small spine; its internal margin dilated and hairy. The internal antennæ are placed very little above the external, and terminate in two short filaments. The external pedipalps are of considerable length, extending forward beyond the peduncles of the external antennæ; the terminal joint much elongated.

The anterior legs are robust and smooth, the hands furnished with a curved moveable finger, which is inflected to meet a small spiniform rudimentary thumb. The remaining legs filiform, elongated; the second the most slender and minutely didactyle; the others monodactyle. A strong spine on the sternum between the anterior pair of legs. Abdomen regularly tapering, rounded, and smooth. The tail with the middle lamina narrow, and pointed at the extremity. Abdominal false feet very long.

Colour greyish-brown, dotted all over with dark brown. Unlike most of its congeners, it does not become red by heat.

Total length from the eyes to the extremity of the tail two inches and a half.

This is one of the most abundant of the coast species of Crustacea. It is taken in multitudes for the table on almost all our sandy shores, ordinarily by means of nets, which are pushed forwards by the "shrimpers," who wade nearly to their middle for hours together, raising the net at intervals, and taking out the shrimps, which are secured in a bag. In some parts of the coast, as at Poole, this species is comparatively rare, and is not used as food. The smaller *Palæmonidæ* are here called "shrimps;" and when of small size and sold by measure, they are termed "cup-shrimps." The present species is called the Sand Shrimp, and the smaller prawns the Rock Shrimp.

In the breeding-season the shrimps approach the estuaries, and even ascend the rivers to a considerable distance.

"Although," says Mr. Thompson, "this species chiefly frequents sandy shores, I have occasionally seen it brought up in the dredge from deep water, and at a considerable distance from land, in the loughs of Strangford and Belfast. Mr. R. Ball mentions that shrimps, though taken in large quantities at Youghal, are held in little esteem; but that the prawn (*Palæmon serratus*), caught abundantly at spring-tides, is much thought of. This latter is called 'shrimp' there, the former the 'gray shrimp;' this term is also used in Smith's 'History of the County of Cork,' written nearly a century since."

The following observations I have selected from the late Mr. Hailstone's MS. Notes on the Crustacea of Hastings. "Although in general this species is very wholesome, yet instances occasionally occur in which it produces effects similar to those which sometimes follow the eating of mussels. They swim in the water or lie upon the sand in shoals, and are taken by a large net with a semicircular mouth, which the shrimper pushes before him along the bottom of the sea during the ebb-tide. In colour they so closely resemble the sand, that, in the pools left by the tide, they are with difficulty distinguished. They are in spawn throughout the year, and cast the shell in March, April, and May."

DECAPODA.
MACROURA.

CRANGONIDÆ.

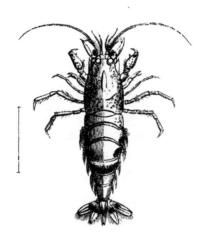

BANDED SHRIMP.

Crangon fasciatus. Risso.

Specific Character.—Second pair of feet shorter than the first and third; abdomen smooth, rather suddenly contracted at the posterior third, with a broad brown band across the fourth abdominal segment: no sternal spine.

Crangon fasciatus, Risso, Crust. de Nice, t. iii. f. 5. p. 82.—Hist. Nat. de l'Eur. Mérid. V. p. 64.— Edw. Hist. des Crust. II. p. 342.

For the first time I am enabled to publish this interesting little species as indigenous to our coasts. I found three specimens amongst some small Crustacea, many of which were very interesting, sent to me by Mr. Alder, by whom they were taken in Salcombe Bay, Devonshire, in the course of his investigations on that coast in search of the nudibranchiate mollusca, the more immediate results of which are well known by the splendid work of that gentleman and Dr. Hancock, published by the Ray Society.

This species considerably resembles the common shrimp in its general aspect; but, besides being very much smaller, it differs from it in many particulars. The peduncle of the internal antennæ is proportionally much shorter; the spines on the branchial region of the carapace obsolete. The first pair of feet are robust, the moveable finger much curved; the second pair of feet shorter than the first and third, extremely small, minutely didactyle; the third pair very slender. The abdomen is as large as the thorax for rather more than half its length, and then contracts somewhat suddenly, by which it may be at once distinguished from young individuals of the other species. There is also a remarkable brown band across the fourth segment of the abdomen, and a spot or two of the same colour on the sides.

Total length six-tenths of an inch.

DECAPODA.
MACROURA.

CRANGONIDÆ.

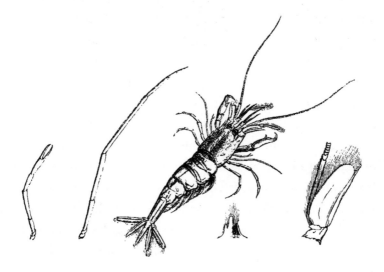

SPINOUS SHRIMP.

Crangon spinosus. Leach.

Specific Character.—Carapace armed with five longitudinal series of teeth; abdomen nearly smooth; the third and fourth segments slightly carinated; the fifth, sixth, and seventh, channelled.

Crangon spinosus,	LEACH, Trans. Lin. Soc. XI. p. 346.—LAM. Hist. Nat. des Anim. sans Verteb. V. p. 202.
Pontophilus „	LEACH, Mol. Brit. t. xxxvii. A.
Crangon cataphractus (in part),	EDW. Hist. des Crust. II. p. 243.

THE carapace of this species is armed with five longitudinal series of teeth directed forwards. The laminar appendage of the external antennæ about the length of the peduncle of the same antennæ. Internal antennæ very short. First pair of feet strong, the moveable finger moderately curved; the second pair extremely small, but distinctly didactyle; the third pair very slender and filiform; the fourth and fifth stronger. The abdomen is nearly smooth; the third and fourth segments obtusely carinated in the centre; the fifth with a triangular depression; the sixth and seventh distinctly channelled.

Length one inch and a half.

A careful examination of several British specimens of this species, and of a well marked one of the Mediterranean form, with which, I believe, it has been erroneously confounded, has led me to reject the alleged synonyms of Risso and Roux, which appear to me to belong to a very distinct species. I am not aware of the grounds upon which Dr. Milne-Edwards has considered the *Egeon loricatus* of Risso as the male of the *Pontophilus spinosus* of Leach; but I feel very confident that they belong to different species.

Although the spinous shrimp is to be ranked amongst the rarer of the small *Macroura* of this country, it is very extensive in its range. Leach speaks but of two specimens known to him; one obtained by Mr. Prideaux in Plymouth Sound, and the other taken off Falmouth by the ill-fated Cranch. Mr. Couch states, in his "Cornish Fauna," that he has obtained it only once, when he found it in the stomach of a fish taken at a depth of from twelve to fifteen fathoms. I have a specimen taken by my friends Professor Forbes and Mr. McAndrew, off Shetland.

By the following extract from Mr. William Thompson's observations on the Crustacea of Ireland, it would appear that it has been found on the Irish coast. "In Mr. V. Thompson's collection there is a specimen bearing the name of *Pontophilus spinosus*, and marked as Irish."

DECAPODA. *CRANGONIDÆ.*
MACROURA.

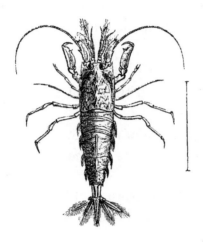

SCULPTURED SHRIMP.

Crangon sculptus. Mihi.

Specific Character.—Carapace with several raised lines, each armed with two or three small teeth; two spines on the median line, one considerably posterior to the other: second pair of legs much shorter than the first, didactyle: abdomen distinctly sculptured; third, fourth, and fifth segments sharply carinated; sixth and seventh, channelled.

OF this new species I possess two specimens, which I found amongst some small Crustacea dredged by Mr. Bowerbank at Weymouth. Like *Egeon loricatus* of Risso, *Pontophilus spinosus* of Leach, and *Crangon septemcarinatus* of Sabine, it has several denticulated carinæ on the carapace, and might at first sight be supposed the young of some one of them. It is, however, a very distinctly marked species.

The carapace is rough, with about five irregular raised lines, each armed with two or three small teeth, and two small spines on the median line, one behind the other. The laminar appendage of the external antennæ, a little longer than their peduncle. The first pair of feet, with the hand, smooth and rounded, the finger moderately curved; the second pair of feet minute. The segments of the abdomen very distinctly sculptured; the raised portions polished, the depressions slightly pubescent: the third, fourth, and fifth segments with a distinct central carina; the sixth and seventh channelled.

Length seven-tenths of an inch.

It differs for *C. spinosus* in the less regularly longitudinal direction of the lateral raised lines on the carapace, the less pointed and fewer teeth of this part, the longer proportion of the antennæ-scale with relation to the peduncle, and, strikingly, in the sculpture of the abdominal segments, and the extent and sharpness of the carina on this part. Of the two specimens which I possess, one has several ova attached to the abdominal false feet.

DECAPODA. *CRANGONIDÆ.*
MACROURA.

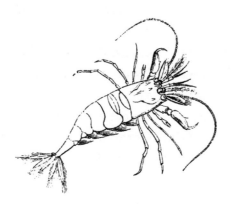

THREE-SPINED SHRIMP.

Crangon trispinosus.

Pontophilus trispinosus,   Hailstone, Mag. of Nat. Hist. VIII. p. 261, fig. 25.

Of this species I have never seen a specimen; I therefore content myself with copying Mr. Hailstone's description, and Mr. Westwood's observations on the species; premising that the characters, as Mr. Westwood very properly remarks, entirely confirm the correctness of Milne-Edwards, and other continental carcinologists, in rejecting the generic separation of *Pontophilus* from *Crangon*.

"On March 1st, 1834, several individuals of a species of *Pontophilus* were brought to me, which had been caught in a shrimping-net upon this coast. They had only three spines on the thorax; one in the middle, and one on each side of it. Their colour was much like a shrimp's, but paler, less clouded, and with a sprinkling of golden blots. Their length about an inch. They were called by the man who caught them 'pug-shrimps;' he said he had never observed them before this last winter. The females were with spawn."

Thus far Mr. Hailstone. Mr. Westwood's observations follow:—

"Of Mr. Hailstone's Crustacea, probably the new *Pontophilus* will be regarded as possessing the highest interest, inasmuch as the propriety of the establishment of the group, which was at first confined to a single species, is thereby proved. The character of this genus, as defined by Dr. Leach, separating it from that of *Crangon* (of which the common edible shrimp, *Crangon vulgaris*, is the type), consisted in the very small size of the second pair of legs, and the length and acuteness of the terminal joints of the external foot-jaws or pedipalpi. These characters, however, to which that of the spinous shell might be added, have been deemed by the French crustaceologists insufficient to warrant the generic separation of the two groups; and, on considering the characters of the new species from Hastings, the correctness of their opinion must, I think, be admitted; since it will be seen, that, in several respects, its characters are quite intermediate between those of the types of *Crangon* and *Pontophilus*. Thus the shell, instead of being armed with a double series of lateral and three rows of dorsal spines, as in the latter, is 3-spinous only, just as in the common shrimp; while the terminal joint of the foot-jaws is scarcely longer than the penultimate joint, and is broad, flat, and obtuse. The central piece of the tail is also much longer than in *Pontophilus spinosus*. Still the minute size of the second pair of legs corresponds with *Pontophilus*; whence it will, perhaps, be more advisable to divide the genus *Crangon* into two sections: first, those with the second and third legs of equal length, the common shrimp; and, secondly, those with the second leg much shorter than the third, the *Pontophili* of Dr. Leach: and even in the former group, the comparatively delicate and imperfect structure of the second pair is very evident; thus proving, in a natural point of view, the generic identity of the two groups."—*Mag. of Nat. Hist.* vol. viii. pp. 261, 265.

DECAPODA.
MACROURA.

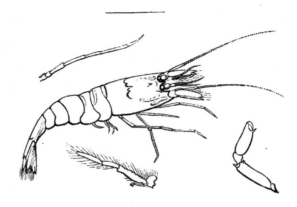

TWO-SPINED SHRIMP.

Crangon bispinosus.

Pontophilus bispinosus, Westwood.
„   „   Hailstone, Mag. Nat. Hist. VIII. pp. 11, 13, f. 30.

The imperfect description and figure of this species, given by Mr. Hailstone, prevent my coming to any very decided opinion as to the distinctness of its specific character. It has, certainly, some points of resemblance to *C. sculptus*, but it is much smaller, if Mr. Hailstone's linear admeasurement be that of an adult; and there is no indication of any sculpture on the abdomen. The following is Mr Hailstone's description:—

"Pedipalps with the last joint rather longer than the preceding one, and bluntish at its termination. First pair of legs compressed, didactyle, with the thumb very short: second pair rather shorter than the first, didactyle; the last joint half the length of the preceding one, which is compressed: third pair very slender, as long as the first pair, with a simple claw: fourth pair and fifth pair of equal length, rather longer than the third, and somewhat thicker but slender; claws simple. Thorax, with two prominent dorsal spines, one considerably behind the other, and on each side a row of blunt notches. A spine at the outer edge of the external plates of the tail."—*Mag. of Nat. Hist.*, vol. viii. p. 273.

Mr. Hailstone states, in his MS. notes, with which he favoured me some time before his lamented death, that one specimen only of this species had come into his possession, which he found at Hastings in a mass of *Filipora filigrana*.

GENUS ALPHEUS. Fabr.

ASTACUS,	Fabr.
PALÆMON,	Oliv.
CRYPTOTHALMUS,	Raffin.
ASPHALIUS,	Roux.
ALPHEUS,	Fabr., Latr., Leach, Edw.

Generic Character.—*External antennæ* placed beneath, and to the outer side of the inner; the lamellar palp of moderate size, sometimes slender and pointed. *Internal antennæ* terminating in two filaments, of which the superior is rather thicker than the inferior; the basal articulation short, and furnished with a spiniform scale. *External pedipalps* more or less slender and elongated; terminal joint broad, and somewhat foliaceous. *First pair of legs* didactyle, robust, one much larger than the other, and very differently formed: *second pair* also didactyle, very slender; the carpus multiarticulate: the remaining pairs slender, monodactyle. *Carapace* extending forwards so as to form an arched covering to the eyes. *Rostrum* small or wanting. *Abdomen* long, and much developed.

Of this genus, which is chiefly confined to hot climates, one species only has been found in Britain, and is now first described.

DECAPODA. GENUS ALPHEUS. ALPHEADÆ.
MACROURA.

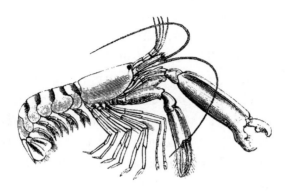

Alpheus ruber.   Edwards.

Alpheus ruber,   EDW. Hist. Crust. II. p. 351.

Specific character.—Rostrum very small. External antennæ without a projecting spine at the base. Arms with a small spine on the upper edge, at a short distance from the extremity. Larger hand with four carinæ, two on the upper surface, and two on the outer; the moveable finger shorter than the other.

THIS is the first instance of any species of *Alpheus* occurring on our coast. The only two specimens hitherto found were obtained by Mr. Cocks of Falmouth, who procured them from the stomachs of cod fish. They are unfortunately much damaged by this circumstance, but enough remains for me to recognise, with scarcely a doubt, its identity with *Alpheus ruber* of Edwards.

The carapace in the specimens is so much injured, that it is impossible to ascertain whether a rostrum existed or not, but the Mediterranean species is stated to possess a small pointed one. The arched processes which always protect the eyes in this genus, and the peculiarities of which form good specific characters, are also destroyed.

The general form of the body is slender. The external antennæ have no spiniform palp at the base, which is also a character of *A. ruber*. The internal antennæ have the general characters of the genus, but the specific distinctions of this part are deficient. The anterior feet are in tolerable preservation. The arm in both is about three times as long as it is broad; and on the upper surface is a small spine, situated about one-third from the extremity: the wrist is very short. The larger hand, which is on the left side, is short, the sides nearly parallel, flattened, the upper margin with two carinæ; the outer side also with two carinæ, the inner surface rounded; the immoveable finger is much curved towards the point, which is acute: there is a distinct and deep depression about the middle of its grasping edge, for the reception of a strong tubercle of the moveable finger; this finger is shorter than the other, and becomes broader towards the extremity, where its outer edge is acutely carinated. The smaller hand is also bicarinated above: the fingers are slender and curved, and there is thus a considerable space between them excepting at the points; their inner margins are hairy. The second pair of feet are filiform, minutely didactyle; the carpus long and multi-articulate. The remaining feet are slender and monodactyle.

Such are the characters afforded by the mutilated specimens before me; and I trust the description, imperfect as it is, may be sufficient to enable some more fortunate observer to compare other and more perfect specimens, and determine with greater precision the identity of this with the Mediterranean species to which I have referred it.

GENUS NIKA. Risso.

PROCESSA.   Leach, Latr.
NIKA.   Risso, Roux, Edwards.

Generic Character.—*External antennæ* much longer than the body, with the basal scale terminating in a single tooth at the outer side; the inner margin hairy. *Internal antennæ* terminating in two filaments, the inner one the longer. *External pedipalps* pediform, with four exserted joints, the second very long, the terminal one pointed. *First pair of legs* dissimilar; that on the right side didactyle, that on the left monodactyle: second pair more slender, very long, filiform, of unequal length; the arm and wrist multi-articulate, the hand minutely didactyle: the remaining pairs very long, filiform, and monodactyle. *Carapace* somewhat elongate, smooth, having a small, simple, compressed rostrum. *Abdomen* a little bent at the third segment; the external laminæ of the tail transversely divided near the extremity.

OF this interesting genus the only known specimen for a long time was a small one on which Leach founded his genus *Processa*, and which was found by Montagu on the southern coast of Devon. Risso, however, had, a short time before Leach's publication, given to the genus the name of *Nika*, of which Leach was not aware at the time. Risso's name must, therefore, be retained, on the ground of priority of publication.

The remarkable peculiarity by which this genus is distinguished from every other form of crustacean, is the dissimilar character of the anterior feet, one being didactyle, and the other monodactyle. In every other instance, however the feet forming a pair may differ in length, in size, or in structure, they agree in the character of the terminal portion, and are always both monodactyle or both didactyle; but in this genus one is invariably and distinctly didactyle, while the other is as distinctly monodactyle.

There are probably several species of this genus in the Mediterranean, and I have in my possession two very distinct species found on our coast, one of which is undoubtedly *Nika edulis* of Risso, and the other an entirely new species.

DECAPODA.   *ALPHEADÆ.*
  MACROURA.

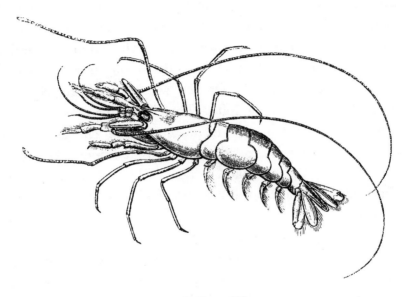

Nika edulis. Risso.

Specific Character.—Didactyle hand longer than the wrist; both straight. Central plate of the tail longitudinally channelled.

Nika edulis,	Risso, Crust. de Nice, p. 85, t. iii. f. 3.—Hist. Nat. de l'Eur. Mérid. V. p. 72.—Roux, Crust. de la Médit. t. xlv.—Lam. Hist. des Anim. sans Vert. V. p. 203.—Desmar. Consider. sur les Crust. p. 230.—Edw. Hist. des Crust. II. p. 364.
Processa canaliculata,	Leach, Malac. Brit. t. xli.
,, *edulis,*	Latr. Reg. Anim. de Cuv. (ed. 2nd) IV. p. 95.
Nika canaliculata.	Desmar Consid. sur les Crust. t. xxxix. f. 4.

The whole of the carapace and abdomen of this species is even and glabrous; the carapace evenly rounded, somewhat compressed, terminating in a short, pointed, slightly carinated rostrum; a short acute point above the outer edge of the orbit. Plate of the external antennæ of almost equal breadth throughout its length; obliquely truncated at the extremity; ciliated on the outer margin. Peduncle of the internal antennæ cylindrical; the internal filament much longer than the external. External pedipalps, with the last joint but two equalling in length the terminal and penultimate joints together. Didactyle foot of the first pair rather thicker than the other; the hand and wrist straight: the monodactyle foot terminating in a very small slightly curved finger. Of the second pair, the right is much longer than the left. The fourth pair longer than the third and fifth, the latter being the shortest. The abdomen is evenly rounded, somewhat compressed, continuous with the carapace. The middle plate of the tail channelled throughout its length, and armed with minute spines.

Length two inches to two inches and a half.

The colour of this beautiful species is described by Risso and Roux as of a flesh-red, more or less dotted with yellow and white, and marked along the back with spots of these colours. The body is said to be so transparent that the viscera may be seen through the integument. The female is stated by Roux to be found, at different periods of the year, bearing eggs of a yellowish-green colour, which are deposited on algæ and fuci.

The same author states that this species constitutes an ordinary article of food on all the coasts of the Mediterranean. It lives in shoals with various species of the genera *Palæmon, Hippolyte, Alpheus,* &c.

After the most careful examination I have been able to institute, I have come to the conclusion that the *Nika edulis* of Risso and those who follow him, and the *Processa canaliculata* of Leach, are identical. It is certainly one of the rarest of our British species. A small specimen was obtained by Montagu at Torcross, on the southern coast of

Devon, and sent by him to Dr. Leach, who founded thereon his genus *Processa*, and figured it in his great work. This specimen is still in the British Museum. That from which my figure and the above description are given was accidentally found by myself in a dish of boiled prawns, on which I was about to breakfast, at Bognor, in the year 1842. Mr. W. Thompson states that there are specimens in the collection made in the south of Ireland by Mr. Vaughan Thompson.

DECAPODA.
MACROURA.

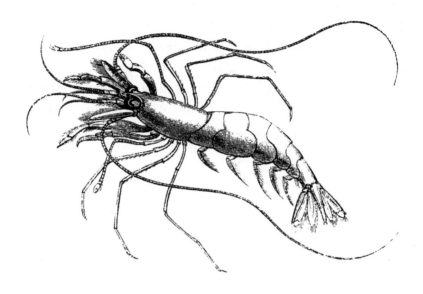

Nika Couchii. Mihi.

Specific Character.—Didactyle hand shorter than the wrist; the former slightly, the latter more considerably curved. Middle plate of the tail attenuated towards the extremity, not furrowed.

The distinctive characters of this new and interesting species are, with the exception of those given above, mostly comparative. The whole animal is longer in proportion to its other dimensions. The carapace, which extends a little further backwards over the posterior thoracic segment at its junction with the abdomen, is longer and more slender; the plates of the external antennæ are longer, and rather tapering towards the extremity. The legs generally are altogether longer and more slender; the didactyle hand and wrist of the first pair are, however, shorter, and both, but particularly the wrist, curved. The fifth pair of legs are quite as long, or a little longer than the third and fourth. The abdomen is notably more slender, and the lateral processes of the segments extend more obliquely backwards; the sixth segment is nearly cylindrical, and the seventh, or middle plate of the tail longer and much narrower, the terminal half being considerably attenuated; the upper surface has no distinct furrow, as in *N. edulis*, and there is on each side a small spine, about the middle of its length, at the point where it becomes narrower. The lateral plates of the tail partake of the general tendency to attenuation, which so remarkably characterises the form of the species.

Length nearly three inches.

One specimen only of this species has come under my notice, and for this I am indebted to Mr. Couch, who sent it to me about five or six years since. It was taken on the coast of Cornwall. I have much pleasure in dedicating so interesting a species to a naturalist who has not only done much for the local Fauna of his own district, but whose observations on the habits and physiology of many forms of marine animals are peculiarly valuable for their truthfulness and originality.

That it is quite distinct from *N. edulis*, the description I have given above sufficiently proves: and it is equally so from any of those, whether varieties or species, named by Risso in his "Crustacés des Environs de Nice." The characters given by this author are, unfortunately, so vague, and the figures in the work just named so bad, that it is frequently impossible to arrive at any tolerable certainty as to the identity of the species described by him, or to ascertain whether the distinctions be specific or not.

DECAPODA.
MACROURA.

ALPHEADÆ.

GENUS ATHANAS. Leach.

Astacus.	Montagu.
Palæmon.	Leach.
Athanas.	Leach, Latr., Roux, Edw.

Generic Character. — *External antennæ* not longer than the body; the scale oval; unidentate on the outer side at the apex. *Internal antennæ* with three filaments, one shorter and thicker than the others. *External pedipalps* short and slender. *First pair of feet* the largest, unequal, didactyle: second pair very small, filiform; the wrist long and multi-articulate; the hand minutely didactyle: the remaining pairs simple. *Carapace* terminating in a simple rostrum. *Abdomen* even; the external plates of the tail transversly divided.

Of this genus one species only is known. It possesses the peculiarity which is found most strongly marked in the Astacidæ, the transverse division of the external lamina of the tail; a character which also obtains in a less marked degree in *Nika*, and some other genera of the Alpheadæ. It differs from all others of the family in having three filaments to the internal antennæ, and from the Palæmonidæ in the simple form of the rostrum and other characters.

DECAPODA.
MACROURA.

ALPHEADÆ.

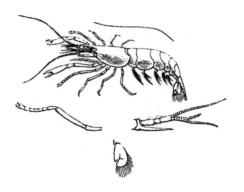

Athanas nitescens.

"*Cancer (Astacus) nitescens*,	Montagu, MSS."—Leach.
Palæmon	,,	Leach, Edin. Encycl. VII. p. 401.
Athanas	,,	Id., p. 432.—Trans. Lin. Soc. XI. p. 349.— Encyc. Brit. Suppl. I. p. 421.—Mal. Brit. t. xliv.—Edw. Hist. des Crust. II. p. 366.

The carapace in this remarkable little creature is very smooth, narrowed anteriorly; the rostrum about half its length, and not extending backwards in a ridge, perfectly simple and pointed. The external antennæ have a cylindrical peduncle about as long as the scale; the latter is nearly oval, hairy on the inner and anterior margin, and with a little tooth on the anterior and outer angle. The internal antennæ have a thick rounded peduncle, terminating in three filaments, the shortest of which is not much more than one-third the length of the longest, and much thicker. The first pair of legs is large and robust in proportion to the size of the animal, and the hand is particularly so: the second pair is very slender; the wrist formed of several joints; the hand very minute and didactyle. The remaining pairs are slender and simple. The abdomen is evenly rounded and smooth. The outer plate of the tail transversely divided about one-third from the extremity. In its general aspect it resembles a very young *Astacus.*

The length of Dr. Leach's specimens is rather more than half an inch.

Colour light buff?

The discovery of this species is due to Montagu, who sent it to Dr. Leach, under the name of " *Cancer* (*Astacus*) *nitescens.*" The latter author states that " it is occasionally found in pools left by the tide amongst the rocks on the coasts of Devon and Cornwall." It cannot, however, be otherwise than rare in that locality, as Mr. Couch does not introduce it in his " Cornish Fauna." Mr .W. Thompson thus records its occurrence on the Irish coast :—" A single specimen was found under a stone, between tide-marks, at Lahinch, county Clare, by Mr. E. Forbes and myself, in July, 1840." These are the only authentic notices of its existence on the British islands that I am acquainted with. Dr. Milne-Edwards mentions its inhabiting the coast of France, but gives no particulars.

The following figure represents one of a series of specimens in the British Museum collection, from Plymouth Sound, remarkable for the large size of the first leg on the right hand side of the body.

GENUS HIPPOLYTE.

CANCER.	O. Fabr.
PALÆMON.	Olivier.
ALPHEUS.	Lamarck, Risso, Sabine.
HIPPOLYTE.	Leach, Desmar., Roux, Edw.

Generic Character. — *External antennæ* placed beneath the internal; the scale externally unidentate. *Internal antennæ* terminating in two filaments: the superior thick, strongly ciliated, and excavated beneath; the inferior slender and setaceous. *External pedipalps* slender and pediform. *First pair of feet* short, equal, didactyle: second pair long, unequal, minutely didactyle; the wrist many-jointed: remaining pairs simple. *Carapace* furnished with a deep rostrum, the carina of which extends over a considerable portion of the median line of the carapace. *Abdomen* abruptly bent downwards at the third segment, which is gibbous, and produced posteriorly.

OF this genus, which was first established by Leach, there are several species inhabiting our coast. They may be at once distinguished from all other *Palæmonidæ* by the peculiar character of one filament of the internal antennæ, which is broad and excavated in its inferior surface. The rostrum is very deep in most species, and the wrist multi-articulate in all. The abrupt curvature of the abdomen varies considerably in the different species.

DECAPODA.
MACROURA.

PALÆMONIDÆ.

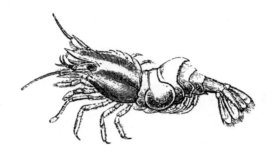

SOWERBY'S HIPPOLYTE.

Hippolyte spinus.

Specific Character.—Rostrum anteriorly truncate, extending backwards nearly to the posterior margin of the carapace, deep, many-toothed above; the teeth, on the portion situated on the carapace, which are three or four, larger than those on the exserted portion. Tooth of the scale of the external antennæ terminal; broad filament of the internal antennæ bent upwards at right angles to the peduncle.

Cancer spinus,	Sowerb. Brit. Misc. t. xxi.
Alpheus „	Leach, Edinb. Encyc. VII. 431.—Trans. of Lin. Soc. XI. p. 347.—Encyc. Brit. Supp. I. 421.
Hippolyte Sowerbæi,	Leach, Malac. Brit. t. xxxix.—Desmar. Consid. sur les Crust. p. 223, t. xxxix. f. 1.—Edw. Hist. des Crust. II. p. 380.

The carapace in this species is furnished with a strong and deep rostrum, the ridge of which rises almost at the posterior edge, and has about four strong and large teeth on that portion which belongs to the carapace, and several smaller ones on the exserted portion, which decrease in size towards the apex; the exserted portion is very deep, terminating in a sharp tooth; the inferior edge with two teeth, of which the anterior is almost as forward as the apex, from which it is separated by a broad notch, which is minutely toothed. There are also two teeth above each orbit, and others on the margin of the carapace. The scale of the external antennæ extends beyond the rostrum, and has a strong tooth on the anterior and outer margin. Terminal filaments of the internal antennæ very short, the thicker one bent abruptly upwards at right angles to its peduncle. Anterior feet not extending beyond the scale of the external antennæ; the hands robust and rounded. The second pair longer than the third, with the wrist divided into about six distinct articulations. The abdomen is very gibbous, the third segment being strongly carinated, the carina terminating in a strong posterior tooth, standing over the middle of the fourth segment. Middle scale of the tail with four pairs of small spines above.

Length about one inch and a half.

I have thought it right to restore the specific name given by Mr. Sowerby to this remarkable species, which Leach adopted on no less than three occasions, and afterwards altered without any sufficient reason. It is the largest of our British species of *Hippolyte*. It is exclusively a northern species, being found, according to Dr. Milne-Edwards, in the seas of Iceland and Greenland. Mr. Sowerby first described it from a specimen found on the Scottish coast: the one figured by Dr. Leach was obtained at Newhaven, in the Frith of Forth. I have two fine and perfect specimens, which were kindly given to me by Mr. McAndrew, who procured them by dredging in deep water off the Isle of Man. It is not yet recorded as having been taken on the Irish coast.

DECAPODA.
MACROURA.

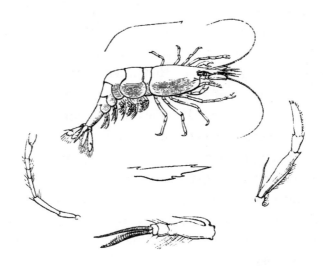

VARYING HIPPOLYTE.

Hippolyte varians.   Leach.

Specific Character.—Rostrum straight, acuminate above, with a spine near the base, and another at the apex; beneath with a sharp two-toothed carina. Antennal scale with the external tooth one-third from the extremity; internal antennæ with the thick filament only slightly curved.

Hippolyte varians,   LEACH, Edinb. Encycl. VIII. p. 432.—Trans. Lin. Soc. XI. p. 347.—Encyc. Brit. Supp. I. p. 421.—Malac. Brit. t. xxxviii. f. 6–16.—EDW. Hist. Crust. II. p. 371.

THE carapace is less gibbous in this than in most other species, although more so than *H. pandaliformis*, now first described; it is terminated by a straight and elongated rostrum, which has on its upper side a tooth near the base, and a very small one near the apex; beneath there is a short carina, which has two teeth: there is also a small tooth on each side of the base of the rostrum, just over the inner edge of the orbit. The scale of the external antennæ is large, and a little longer than the rostrum; its external tooth is placed at the distance of about one-third from the extremity. The thicker filament of the internal antennæ is of moderate size, and is but slightly curved, instead of being abruptly bent at right angles, as in *H. spinus* and some other species. The external pedipalps are of moderate length; the terminal joint short, flattened, rounded, hairy, and furnished with minute spines on its inner margin. The first pair of feet very short, rather thick; the second pair shorter than the third, and the wrist with not more than three or four joints. Abdomen less gibbous than in some species, as *H. Spinus* and *H. Cranchii*. Middle plate of the tail with two pairs of small spines.

Length about three-quarters of an inch.

The usual colour is a beautiful clear green; but, as Dr. Leach states, "it is very variable in colour, occurring with every shade of green, and of every tint between reddish and liver-brown."

This is the most abundant of all our species of *Hippolyte*, though probably not the most extensively distributed. "It is found," says Dr. Leach, "in profusion in pools amongst the rocks on the south-western coast of Devon and Cornwall." It is common all along that coast, and as far as Poole Harbour in Dorsetshire; and, although it is not mentioned by Mr. Couch in his Cornish Fauna, I have received specimens of it from that gentleman from Polperro. It has been found extensively round the Irish coast. Mr. W. Thompson says, "It has been taken commonly by Mr. Hyndman and myself in the rock pools accessible at low-water throughout the Down coast, and has been dredged by us in deep water on the north-east coast, and in Killery Bay, Connemara. Mr. R. Ball has specimens from the shores about Dublin."

It is a beautiful and elegant species, but loses its lovely green colour soon after death.

DECAPODA. *PALÆMONIDÆ.*
MACROURA.

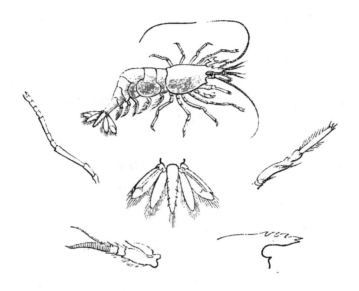

CRANCH'S HIPPOLYTE.

Hippolyte Cranchii. Leach.

Specific Character.—Rostrum short, incurved at the base, with three teeth above, the apex emarginate, bidentate, the upper tooth the longer; beneath unarmed.

Hippolyte Cranchii,   Leach, Malac. Brit. t. xxxviii. f. 17–21.—Edw. Hist. Crust. II. p. 376.—Couch, Corn. Faun.

The carapace is short and rounded; the rostrum short, raised, and somewhat abruptly incurved at the base, where it is broad, and armed with three conspicuous teeth; the apical portion straight, bifid at the extremity, the lower tooth shorter than the upper; the inferior edge is short, and without any tooth; there is no tooth above the orbit. The scale of the external antennæ extends to more than half the length of the filaments of the internal, and the marginal tooth is terminal. Internal antennæ with the thicker filament very slightly curved. Anterior feet extending forwards a little beyond the antennal scale; second pair with the wrist long, and formed of six articulations. The junction of the thorax and abdomen is very gibbous, the process on the posterior margin of the third segment rounded, and but little prominent. The middle portion of the tail has four pairs of extremely minute teeth, so small as to be discerned with difficulty.

Length about three-quarters of an inch.

This little species, which is about the size of *A. varians*, may be at once distinguished from it by the thicker thorax, the more gibbous abdomen, the strong line of demarcation between those two parts, and especially by the form of the rostrum. It is a widely extended species, and in some parts abundant. It was taken first by Mr. Cranch, and afterwards by Mr. Prideaux, in the Kingsbridge estuary; it is also admitted into Mr. Couch's Cornish Fauna. I have received it from Torbay, and from Salcombe Bay, through the kindness of Mrs. Griffiths and Mr. Alder: it was dredged at Poole by my relative, Mr. Henry Salter; and I have specimens taken by Professor Forbes and Mr. M'Andrew in Loch Fyne. Mr. Wm. Thompson gives it as an Irish species, only on the authority of a specimen in the collection of Mr. Vaughan Thompson: it is exceedingly probable, however, that it will be again found on the coasts of Ireland, as it has so extensive a range on those of England and Scotland.

DECAPODA.
MACROURA.

THOMPSON'S HIPPOLYTE.

Hippolyte Thompsoni. Mihi.

Specific Character.—Rostrum straight, deep, acute, continuous with a slight carina, which extends from near the posterior margin of the carapace; furnished above with eight teeth, of which four, more distant than the others, are situated on the carapace; beneath, with three minute teeth near the apex.

The carapace is of moderate length, the surface slightly scabrous: a carina commences at about two-thirds backwards, and extends forwards to form the upper portion of the rostrum, which, with the carina, is furnished with eight teeth, directed forwards, of which four are situated on the carapace, and are more distant from each other than those on the rostrum itself; the inferior portion of the rostrum has three minute teeth placed towards the apex, which is acute and simple. The scale of the external antennæ extends to more than half the length of the internal, and the latter have the filaments slightly curved downwards, and moderately thick. The anterior pair of feet are of moderate length; the hand thicker than the wrist, which is rather short, and has but few articulations. The abdomen is remarkably gibbous, and the anterior segments are very slightly scabrous at the sides.

Length nearly an inch.

Of this new species of *Hippolyte* I have seen but one specimen,—a female loaded with extruded ova, which I received from my friend Mr. W. Thompson of Belfast, the acute and successful investigator of the zoological treasures of his own country, by whose labours our knowledge of the natural history of that part of the United Kingdom has been so much enriched, and to whose name I dedicate the species. It was obtained by that gentleman, with specimens of *H. Cranchii*, on the north-west coast of Ireland.

It differs from *H. pandaliformis* in the form of the rostrum, which is shorter, and toothed throughout its whole length, as well as in its general figure, which in the present species has more of the normal aspect of the genus, and less resemblance to the more typical *Palæmonidæ*. From *H. Cranchii* it differs in the longer, straighter, and more toothed rostrum, and in the less gibbous form of the thorax and of the abdomen. With no other species could it be confounded, even at the first glance.

DECAPODA.
MACROURA.

PALÆMONIDÆ.

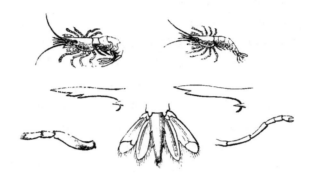

PRIDEAUX'S HIPPOLYTE.

Hippolyte Prideauxiana. Leach.

Specific Character.—Rostrum quite straight, acuminate, above unarmed, beneath with one or two teeth on the anterior portion.

Hippolyte Prideauxiana,   LEACH, Mal. Brit. t. xxviii. f. 1. 3, 4, 5.
„   „   *var.,*   EDW. Hist. des. Crust. II. p. 372.
„   *Moorii,*   LEACH, l. c. f. 2—EDW. l. c.

THIS little species very much resembles *H. varians*; it differs from that species, however, in the form of the rostrum, which is of more equal breadth throughout the principal part of its length, in the absence of any tooth on its upper side and of the deep carina on the lower.

I have thought it right to include the animals defined by Dr. Leach under the above specific names, as one species, being unable to find any distinctive characters of sufficient importance to warrant their separation. I have, therefore, taken *Prideauxiana* as the normal form, and as having the priority in nomenclature; and have given

Moorii merely as a variety—a view which Dr. Edwards had already taken in his "History of Crustacea," although it still remains to be proved which of the two forms of rostrum is, on account of the comparative frequency of occurrence, to be considered as normal.

The wrist of the second pair of legs has only two short and one long articulation; the abdomen is remarkably bent at the third segment. The whole animal is smaller than *H. varians*, and of a reddish brown colour.

This is certainly a very rare species, and is not mentioned in Mr. Thompson's Irish Fauna, nor in that of Cornwall by Mr. Couch. I have received it, however, from the neighbouring coast of Devonshire.

The vignette below appears to be a representation of the Common Shrimp (*Crangon vulgaris*), and was taken from a tesselated pavement discovered at Cirencester in 1783. (*Vetust. Mon.* vol. ii.)

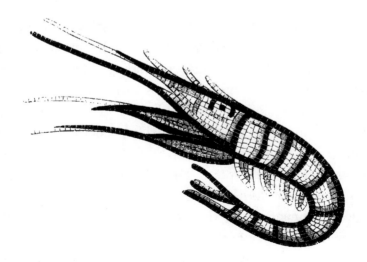

DECAPODA. PALÆMONIDÆ.
MACROURA.

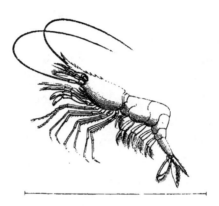

Hippolyte pandaliformis.   Mihi.

Specific Character.—Rostrum extending beyond the scale of the antennæ, nearly straight, slightly turned upwards, with seven teeth on the upper and three on the lower edge; thicker filament of the internal antennæ moderately curved.

THE carapace in this species is evenly rounded, with a slight carina on the anterior third, passing into the rostrum, which is nearly straight, but a little turned upwards towards the extremity, and extending beyond the antennal scale; it is furnished above with seven acute teeth, of which three are on the carina of the carapace, and the remainder on the free portion; beneath are three similar teeth, and the apex is bifid, the upper point projecting beyond the lower. The eyes are remarkably large, as in *Pandalus annulicornis*. The external antennæ have the long filament longer than the body; the scale is rather narrow, particularly anteriorly, with long cilia on the outer edge, and externally a small tooth near the extremity. The thick filament of the internal antennæ is of moderate size and slightly curved; the external pedipalps slender and pediform; the anterior pair of feet about two-thirds the length of the second; the wrist of the latter six-jointed, the third joint about twice the length of each of the others; the remaining legs long and slender. The abdomen is more slender than in any other species of *Hippolyte* with which I am acquainted; the terminal joint narrow and acutely pointed.

Total length an inch and a half.

The resemblance of this species to *Pandalus annulicornis* is so remarkable, that I have given it a specific name in accordance with that relation, in order to record its probable situation as leading from the normal forms of the genus towards *Pandalus*. It is strikingly abnormal as regards its own genus. The general slenderness of the whole body, the even form of the carapace, the length and form of the rostrum, the length of the legs and antennæ, all exhibit a marked tendency in the direction alluded to. It is, however, in all its essential characters, a true *Hippolyte*.

This constitutes another interesting addition to our native Crustacea, which we owe to the labours of Mr. M'Andrew and Professor Forbes, by whom it was dredged in Loch Fyne, at a depth of about twenty fathoms.

I have two specimens, for which I am indebted to those gentlemen, and this is the only instance in which its occurrence has been recorded.

GENUS PANDALUS, Leach.

Astacus.	Fabr.
Palæmon.	Risso.
Pontophilus.	Id.
Pandalus.	Leach, Latr., Edw.

Generic character.—*External antennæ* longer than the body, the antennal scale unidentate on the outer margin. *Internal antennæ* with two filaments, the external one thicker; basal joint of the peduncle hollowed above for the lodgment of the eyes. *External pedipalps* slender, pediform. *First pair of feet* slender, shorter than the others; the terminal joint styliform, simple; *second pair* filiform, didactyle, unequal; one much longer and more slender than the other, both with the wrist and arm multiarticulate; *third, fourth, and fifth pairs* of feet, slender, slightly diminishing in length. *Carapace* armed with a long rostrum, the carina of which extends half way to the posterior margin of the carapace, the rostrum curved upwards, and denticulate above and below. *Abdomen* with the third joint gibbous, the upper and posterior margin produced backwards.

I have always considered this genus as affording a distinct passage from the genus *Hippolyte* to *Palæmon*; a view of its relations, which has received an important confirmation in the discovery of the *H. pandaliformis*, which may be considered as the osculant species on that side. Like *Hippolyte* it has but two filaments to the internal antennæ and the carpus multi-articulate; so that, in fact, it resembles that genus in its essential characters even more nearly than *Palæmon*.

DECAPODA.　　　　　　　　　　　　　　　　*PALÆMONIDÆ.*
MACROURA.

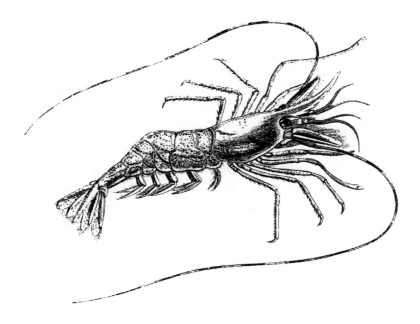

Pandalus annulicornis. Leach.

Specific Character.—Rostrum the length of the carapace; anterior half without teeth above, excepting a small one close to the apex.

Pandalus annulicornis.　　Leach, Malac. Brit. t. xl.—Latr. Encycl. Meth.—
　　　　　　　　　　　　Lam. Anim. sans vert. V. p. 203.—Edw. Hist. des
　　　　　　　　　　　　Crust. II. p. 384.

The carapace of *Pandalus annulicornis* resembles strongly that of the genus *Palæmon.* The rostrum, as in that genus, is very long; it is considerably turned upwards towards its extremity; the carina commences about half-way towards the posterior margin of the carapace, and it is finely toothed to nearly half of the free portion of the rostrum, the remainder being without teeth above, excepting a minute one just above the apex; beneath, it has five distinct teeth. The eyes are remarkably large. The external antennæ have very long filaments, which are marked with alternate rings of dark and light colour through their whole length; the base cylindrical, and the scale diminishing in breadth forwards, with a small tooth on the exterior margin at the extremity. The internal antennæ have but two filaments, the external of which is the thicker; the base is hollowed to receive the eyes, and there is, at the anterior margin of this excavation, a fringe of hairs, which covers the inferior part of the eyes, and affords them protection. The external pedipalps have the basal joint hollowed above the terminal joint, furnished with stiff hairs and small spines. The anterior feet are simple, slightly curved, the basal joint cylindrical, the remainder styliform, and acute at the termination. The second pair of feet are of very unequal length and size; one being very slender, very long, the wrist and arm multi-articulate, the didactyle hand very minute; the other thicker, shorter, likewise didactyle, and with the arm and wrist multi-articulate; the remaining feet nearly of equal length, and simple; the terminal joint furnished with a row of spines beneath. The abdomen resembles that of *Hippolyte* in the gibbous form of the third segment; the centre piece of the tail has three pairs of small teeth on its anterior half.

The usual length is from two inches to two inches and a half.

It is of a reddish grey colour, curiously dotted and marked with deeper red.

At first sight this species may be readily mistaken for a common prawn; but a closer examination will shew that its structural relations are much nearer to *Hippolyte* than to *Palæmon*. Its distinction from the prawn appears to have struck several persons about the same time. It was first discovered, according to Dr. Leach, by the

Rev. Dr. Fleming, in Zetland, and in St. Andrew's Bay, Scotland; it was also observed by Montagu on the coast of Devon; and by Mrs. Dawson Turner, who noticed it at Yarmouth, and pointed it out to Mr. J. D. C. Sowerby as distinct from the common prawn. " It is used," says Dr. Leach, " at Yarmouth as an article of food; and is at that place so much esteemed for the table, as to afford constant employment during the summer season to several fishermen, who take it in abundance at a considerable distance from the shore, and name it from that circumstance the sea-shrimp."

The extent of the range of this species is very remarkable. Dr. Milne Edwards mentions its being an inhabitant of Iceland. We see above that Dr. Fleming obtained it at Zetland; I have specimens taken by Mr. M'Andrew and Professor Forbes about the same locality; and it is found commonly on the southern coasts of England. Mr. Couch admits it into his Cornish Fauna, and gives it the expressive name of " Æsop-shrimp " (another proof of its affinity to *Hippolyte*). I have specimens from Poole Harbour, in Dorsetshire, and on the Norfolk coast it is constantly taken and sold as a " prawn." I have occasionally known it brought to the London markets, where, however, it is usually seen of small size. As an Irish species, it is stated to occur in Mr. J. Vaughan Thompson's collection; and Mr. William Thompson adds,—" It has been taken commonly by Mr. Hyndman and myself in the rock pools accessible at low water throughout the Down coast, and has been dredged by us in deep water on the north-east coast, and in Killery Bay, Connemara. Mr. R. Ball has specimens from the shores about Dublin."

DECAPODA. PALÆMONIDÆ.
MACROURA.

GENUS PALÆMON. Fabr.

ASTACUS. Pennant.
PALÆMON. Fabr., Latr., Lam., Leach, Edw.

Generic character.—*External antennæ* placed beneath, and a little to the outer side of the internal; the lamelliform palp very large, nearly oval, rounded and ciliate at the apex and armed with a spine near the extremity of the outer margin. *Internal antennæ* inserted above the external; the first joint of the peduncle very large, depressed, excavated on its upper side to receive the eyes, and armed on the outer side with a strong spine; the two following joints large and cylindrical, the last bearing three setæ of which two are very long, and the other very short and curved. *External pedipalps* of moderate length, pediform, slender, terminating in a slightly curved nail. *First pair of feet* very small and slender; hand didactyle: *second pair* also didactyle, much larger than the former; the remaining pairs simple, monodactyle. *Carapace* of moderate size, broad, terminating in a long, laterally flattened rostrum, which extends usually beyond the peduncles of the antennæ. *Eyes* large and projecting. *Abdomen* large, diminishing regularly towards the tail, and rounded on the upper surface; the terminal segment, which forms the middle portion of the tail, triangular. *Abdominal false feet* very large; those of the first pair furnished with a large ciliated scale, and a much smaller one; the others with two scales, which are also distinctly ciliated.

ALTHOUGH most of the species of this genus are of moderate size, there are some inhabiting the tropical regions which may almost rival the larger *Astacidæ*. The

Palæmon Carcinus, for instance, sometimes reaches to nearly a foot in length, and *P. Jamaicensis* is nearly as large. The common prawn, *P. serratus*, is the species best known and most esteemed in our climate; but a very careful examination of all the means within my reach, has established in my mind, very satisfactorily, the existence of four distinct British species of *Palæmon*, forming one additional to those before defined.

DECAPODA. *PALÆMONIDÆ.*
MACROURA.

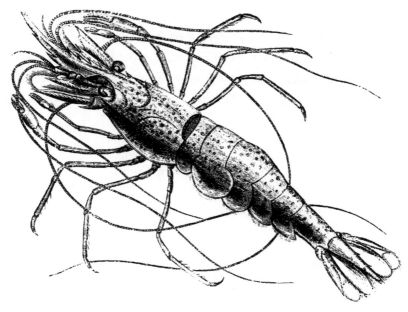

COMMON PRAWN.

Palæmon serratus.

Specific Character.—Rostrum extending considerably beyond the antennal scale; turned upwards anteriorly; bifid at the extremity; above, armed with seven or eight teeth, the anterior third unarmed; beneath, with five or six teeth.

Astacus serratus,	PENN. Brit. Zool. IV. t. xvi. f. 28, p. 19.
Cancer squilla,	HERBST. II. t. xxv. f. 1.
Palæmon ,,	LATR. Gen. Crust. I. p. 54.—LEACH, Edin. Encycl. VII. 401.
,, *serratus.*	FABR Supp. p. 604.—LEACH, Trans. Linn. Soc. XI. 348.—Mal. Brit. XXIV. t. xliii. f. 1-10.—EDW. Hist. Nat. des Crust. II. p. 389.

THE carapace of the common prawn is even, rounded, and furnished anteriorly with two points, one above and the other beneath the peduncle of the external antennæ; the rostrum is of great length, the anterior half ascending;

above armed with seven or eight teeth (usually seven, rarely six), which are confined to the posterior portion, the anterior third being slender and unarmed; the extremity bifid, the inferior point being the longer; beneath armed with four or five teeth (usually five). The eyes are large and round. Of the three filaments of the internal (superior) antennæ, the shortest scarcely extends to the extremity of the rostrum; the others are more than twice as long. The external antennæ are very long, being half as long again as the animal from the tail to the extremity of the rostrum; the scale with the sides nearly parallel, anteriorly and posteriorly obliquely truncate, forming a long rhomboid; the inner edge furnished with long hairs. The first pair of feet very slender, ordinarily bent upon itself; the hand and fingers together not nearly as long as the wrist; the second pair extend forwards to the end of the rostrum; the hand rounded, elongate; the fingers slender, as long as the hand; the hand and fingers together twice as long as the wrist; the remaining pairs slender and simple. The abdominal false feet very long; the terminal joint of the abdomen narrowed forwards, with two long slender terminal teeth, and two pairs of small teeth on the sides. Caudal laminæ furnished with long hairs on the terminal margin.

Ordinary length upwards of four inches.

Colour bright grey, spotted and lined with darker purplish grey.

This species, which is so well known as a favourite and delicate article of food, is found in vast numbers on all the coasts of this island. It appears from various accounts that it approaches the shore in its young state, and multitudes of them are taken in shrimp-nets, and sold as shrimps on some parts of the coast. I found that at Bognor the fishermen consider them, when young, as a distinct species, and assert that, at certain seasons, they drive the true *prawns* from their ordinary place of resort. The probability is that at the season when the young ones have arrived at a certain size, they separate themselves from the older ones, which at that period of the year retire further from the shore. At Poole I have found the young ones of this species associated with two other species of *Palæmon*, and the three are ordinarily sold there under the name of " cup-shrimps," being measured in small cups, instead of being sold by tale, as they are when larger. When of middle size they still retain the name of shrimps at that place, and are only called prawns when they acquire larger dimensions.

In the adult condition they frequent rocky parts of the coast, delighting in still transparent water, where they may be seen in numerous companies, disporting, in the most elegant and beautiful manner, amongst the long fuci which wave in the tide.

Prawns are sometimes taken in bag-nets, suspended from a circular ring of iron, at the end of a pole; but in many parts, the finest are caught in pots, resembling lobster-pots, but smaller, and made of a closer fabric. At Bognor I found that besides the baited pots or traps, each fisherman had a store-pot, into which he transferred the prawns, when he went his round of the baited pots, and kept them there until they were wanted.

They are found with the ovaries filled with spawn, and with the abdominal false feet loaded with excluded spawn at all seasons of the year. They are chiefly obtained for the London markets off the Isle of Wight and Hampshire coast; but they are much deteriorated by the time which elapses after their capture, before they can be procured in the metropolis.

DECAPODA. *PALÆMONIDÆ.*
MACROURA.

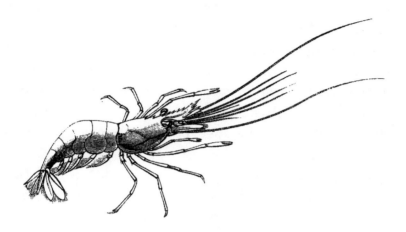

Palæmon Squilla. Fabr.

Specific Character.—Rostrum nearly straight; the apex emarginate; above with seven or eight teeth, of which two are on the carapace, and the third immediately above the ocular notch; beneath with three teeth.

Cancer Squilla,	Linn. Syst. Nat. I. 1051.
Astacus „	Fabr. Ent. Syst. II. p. 485.
Palæmon „	Fabr. Suppl. Ent. Syst. p. 403.—Latr. Hist. des Crust. et des Ins. VI. p. 257.—Leach, Edinb. Enc. VII. p. 432.—Malac. Brit. t. xliii. f. 11-13.—Edw. Hist. des Crust. IV. p. 390.—Couch, Corn. Faun. p. 80.—W. Thompson, Crust. of Ireland.

This species differs from the former in a few distinct, but, with one exception, not very tangible characters. The whole animal is much smaller, being not more than half the length; the first pair of feet are shorter in proportion, and the second pair less robust. But it is in the rostrum that the principal and most obvious distinctive characters exist. This part is almost straight, having, however, a very slight curve upwards towards the extremity; it has seven or eight teeth on the upper side, and three on the under; but the number alone, although very constant, scarcely constitutes so true and certain a criterion as the fact that of the upper teeth two are invariably placed on the median line of the carapace, posterior to the base of the rostrum, and the third immediately over the margin of the ocular notch. The upper teeth are very acute, spiniform, and directed very much forwards. Those of the under side are broader at the base, triangular, and the posterior one slightly falcate. The apex of the rostrum is bifid, the inferior point being the longer.

Of fourteen specimens examined I found the teeth on the upper and under side of the rostrum as follows:—Seven had $\frac{8}{3}$, five $\frac{7}{3}$, one $\frac{9}{3}$, and one $\frac{7}{4}$, so that the normal number is $\frac{7}{3}$ or $\frac{8}{3}$.

The total length, from the rostrum to the tail inclusive, of the largest specimens I have examined, was two inches one line.

This species is pretty widely distributed along our coasts. I have obtained it from Ireland through the kindness of Colonel Portlock; and Mr. W. Thompson records it as common on the shore of Belfast Lough, in rock pools, on the Down coast, as well as in deep water. He also mentions having met with it commonly in rock pools about Ballantrae, Ayrshire. It occurs on the Cornish and Devonshire coast, but Mr. Couch considers it rare in the former county, although Dr. Leach mentions it as very abundant in the latter. At Poole, in Dorsetshire, it forms a considerable proportion of the " cup-shrimps," a name given there to the young prawns of three different species, which are sold by measure.

DECAPODA.　　　　　　　　　　　　　PALÆMONIDÆ.
MACROURA.

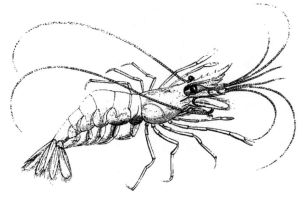

Palæmon Leachii.

Specific character.—Rostrum nearly straight, with five or six teeth above, and three beneath; one only of the former situated behind the line of the ocular notch; apex generally emarginate.

Amongst the smaller *Palæmonidæ* found in Poole Harbour, to which I have already alluded as being sold there under the name of "cup-shrimps," there are found a considerable number which differ materially in the form of the rostrum, as well as in the number of teeth with which it is furnished, from either of the species hitherto described. I have thought right to describe it as a distinct species, which I cannot doubt to be correct. It differs from *P. Squilla* in the smaller number of the teeth in the upper crest of the rostrum, and in the fact that one of those teeth only is placed posterior to the ocular notch. From *P. varians* it is more obviously distinct.

Of twenty specimens taken promiscuously, the number of teeth on the upper and under edges of the rostrum were as follows:—ten had $\frac{5}{3}$, seven $\frac{6}{3}$, two $\frac{6}{4}$, and one had only two teeth beneath. The normal number, therefore, may be considered as $\frac{5}{3}$.

It may perhaps appear, without a careful comparison of numerous specimens of both species, that the characters above named are scarcely sufficient to warrant their separation; but it happens that an additional distinctive character obtains invariably, associated with the different number of the rostral teeth. The rostrum in the present species is covered with innumerable reddish dots, which continue to be very visible from their opacity, even after all colour has been removed from long immersion in spirit. In *P. Squilla* there are no such coloured dots.

I have given to this species the name of my lamented friend, the distinguished illustrator of British Carcinology, the late William Elford Leach.

DECAPODA.
MACROURA

PALÆMONIDÆ

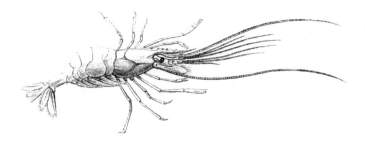

Palæmon Varians. Leach.

Specific character.—Rostrum perfectly straight, the apex entire; above with four to six teeth, beneath with two. Scale of the external antennæ rounded at the apex.

Palæmon varians,   Leach, Edinb. Enc. VII. p. 401. 431.—Id. Trans. Lin. Soc. XI. p. 349.—Id. Malac. Brit. t. xliii. f. 14-16.—Edw. Hist. des Crust. IV. p. 391.

The absolute restriction to two teeth on the inferior crest of the rostrum in this species, would distinguish it from all other British species of *Palæmon*, even without the additional characters of the small number of teeth on the upper crest, and the entire apex. The whole rostrum is perfectly straight, lanceolate, acute at the apex, as long as the scale of the external antennæ; one of the teeth on the upper crest is always placed a little behind the ocular notch. These teeth are generally four in number, sometimes five, very rarely six; beneath there are never more than two teeth; I have seen one specimen in which there was only one. The external antennæ are of moderate length, not much exceeding that of the body; the scale anteriorly rounded, in which character it differs from *P. serratus*. The hand of the anterior pair of feet is slightly tumid; the upper finger hairy.

This species is less widely distributed, as far as we can at present judge, than either of the former ones. It is, however, found on the Devonshire and Dorsetshire coast, and onwards as far as that of Norfolk. Mr. Couch does not include it in his Cornish Fauna. Mr. Thompson states that a few specimens have been taken in Belfast and Strangford Loughs, and I have received it myself from Ireland through the kindness of Col. Portlock.

The vignette consists of several variations of the rostrum in three species of Palæmon. The first three on the left are of *P. Squilla*. The three in the middle belong to *P. serratus*, and the three on the right to *P. Leachii*.

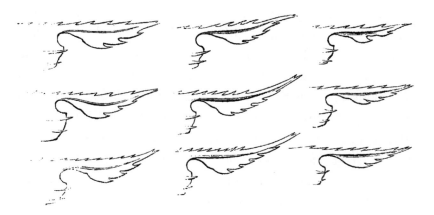

GENUS PASIPHÆA. Sav.

ALPHEUS.   Risso.
PASIPHÆA.   Savigny, Risso, Desmar. Latr. Leach, Edw.

Generic character.—*External antennæ* placed beneath the internal; the basal joint of the peduncle narrower than the succeeding one. *Internal antennæ* with the peduncle slender, and terminated by two filaments, one of which is considerably longer than the other. *External pedipalps* very long, slender, and pediform; furnished at the base with a lamellar ciliated palp. *First and second pairs of legs* didactyle, rather robust, nearly of equal length; the fingers slender and curved; *third, fourth, and fifth pairs* very slender, monodactyle, the fourth pair the shortest. *Carapace* very much elongated, compressed, narrowed anteriorly. *Abdomen* very long, much compressed; the fifth segment broad and squared at the lateral margin; the sixth very long; the seventh narrow and wedge-shaped; *the false feet* of the first segment with the filaments rudimentary, the others with two equal.

The history of this remarkable genus has hitherto been involved in some obscurity. In the description of our species I have given some reasons for believing that as yet one species only is known.

DECAPODA.
MACROURA.

PENÆADÆ.

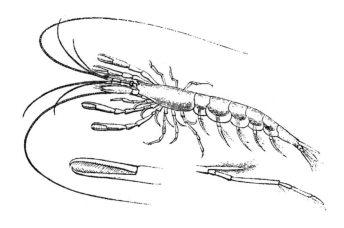

Pasiphæa Sivado, Risso.

Specific character.— External caudal laminæ longer than the internal, which are longer than the middle one.

? *Alpheus Sivado,*	Risso. Crust. de Nice, t. iii. f. 4. p. 93.—Desmar. Consid. sur les Crust. p. 240.—Latr. Regn. Anim. de Cuv. IV. p. 99.	
? *Pasiphæa*	„	Risso, Hist. Nat. de l'Eur. Merid. V. p. 81.—Edw. Hist. Nat. des Crust. IV. p. 426.— Guer. Iconog. du Reg. Anim. (Crust.) t. xxii. f. 3.—Thompson, Ann. Nat. Hist. V. p. 256.
„	*Savignii,*	Leach, MSS. in Mus. Brit.—Edw. l. c. p. 426.
„	*brevirostris,*	Edw. l. c. p. 426.

The general form of this remarkable species distinguishes it at first sight from every other known Crustacean. The whole body is exceedingly compressed laterally, and the carapace elongated and somewhat attenuated forwards. There is scarcely a perceptible rostrum; but immediately behind the anterior margin of the carapace, is a small triangular tooth, with the point turned forwards, and a similar one, but smaller, on each side just above the origin of the superior or internal antennæ. The superior antennæ have a cylindrical peduncle, the basal joint of which is hollowed for the lodgement of the eyes; the filament is double. The external or inferior antennæ are placed immediately beneath the former; the peduncle is cylindrical, terminating in two filaments, one of which is about as long as the body, the other extremely short; the lamina or scale is narrow-ovate, and ciliated along the inner margin. The eyes are much larger than their peduncles. The external pedipalps are long, slender, and pediform, each furnished with a long palp which is a little thicker than it and about half its length. The first pair of feet are rather more robust than the second, the hand thicker in the middle, the fingers shorter than the hand, and curved at the points which are acute, and cross each other when closed; the arm is furnished with a series of short remote teeth; the second pair are ordinarily rather longer than the first, more slender, the fingers as long as the hand, and furnished along their prehensile edge with a dense series of short, stiff hairs. The third pair is extremely slender, filiform, and simple; the fourth pair, by much the shortest, being not more than half the length of the fifth, the penultimate joint furnished on its inner margin with a brush of stiff hairs, and the terminal joint ciliated; the fifth pair very long and slender, the terminal joint ovate, furnished with a lash of hairs of twice its length. The abdomen is very long and much compressed; the second segment broad and rounded; the fifth broad and squared at the lateral margin; the sixth remarkably long and narrowed; the seventh or central lamina of the tail very narrow and wedge-shaped. The external laminæ of the tail are longer than the internal, which are intermediate in length between the former and the median lamina or seventh abdominal segment.

The abdominal false feet have two equal filaments, with the exception of the first pair, of which one of the filaments is extremely small or rudimentary. The eggs are remarkably large and not numerous.

The colour of the Mediterranean species described by Risso, and which I believe to be identical with this, is thus given by that author:—The body is white, slightly iridescent, transparent, banded with red at every articulation; the eyes black; the antennæ, pedipalps, and feet red, and the caudal scales dotted with the same colour.

Total length of specimens from the Bristol Channel about three inches.

I have already alluded to the obscurity in which this genus has been involved; and which has arisen, in great measure, from the extremely erroneous figure given by Risso in his " Histoire des Crustacés de Nice," &c. This figure in fact is so bad, that it affords no ground whatever for any determination of the species: it is indeed much to be regretted that in a work in which so many interesting species were first described, the figures are almost universally so imperfect as to afford no specific character which can be at all depended upon. Savigny, in his masterly " Memoires sur les Animaux sans Vertébres," establishes the genus by name, but without any description, retaining the specific name of *Sivado* after Risso. Leach, who appears from some other circumstances to have been unacquainted with Risso's earlier work, gives to a specimen in the British Museum the name of *P. Savignii*, but he has not, as far as I am aware, published any account of it. Milne Edwards, upon the credit of Risso's figure, has considered the Mediterranean species as distinct from the British; and he has added a third species, which he calls *P. brevirostris*. I do not think, however, that the characters by which this eminent carcinologist distinguishes the latter supposed species can be considered as constant, as I have seen British specimens which vary considerably in the degree of ciliation in the parts from which he deduces his distinctions. It is, finally, very evident that the figure of Guérin in his " Monographie," imperfect and unsatisfactory as it is, and which he refers to *P. Sivado,* belongs to our British species; and the result of all these considerations in my mind, is the full conviction that we are at present acquainted with but one species of this genus: I have therefore retained Risso's original specific name of *Sivado.* It now remains that I should state what is known about this species as a native of Britain. It appears probable, from the following extract of a letter from Dr. Leach to Mr. Baker of Bridgewater, that the first British specimen known was in the collection of Mr. Sowerby; and it also appears from the same passage that Mr. Baker had himself sent another individual of this species to Dr. Leach with some other crustacea and insects. Dr. Leach writes, " I cannot refrain from noticing two species which give me the most pleasure. The one is a species of the genus *Pasiphæa* of Savigny. I have seen a specimen of this genus in Mr. Sowerby's collection, and I believe it to be the same species." Subsequently Mr. Baker obtained several others, taken, I believe, in the Bristol Channel, which are now, through his kindness, in my possession, and I have lately received from Mr. M'Andrew, two individuals, a male and a female, taken by him in the Irish Channel, the latter having the ova excluded and attached to the abdominal false feet. It would appear by the following notice in the fifth vol. of the " Annals of Natural History," by Mr. W. Thompson, that the specimen in the British Museum was originally taken on the coast of Ireland. "*Pasiphæa Sivado*. In the British Museum there is a specimen so named and labelled 'Ireland.' From the donor, the Rev. James Bulwer, I learned that it was taken by him in the vicinity of Dublin."

I have, for obvious reasons, dwelt more at large than in a case of less difficulty would have been necessary, on the characters as well as the nomenclature of this species. It is now for others, who may have the opportunity, to compare Mediterranean with British specimens, in order to ascertain whether I am right in considering them as all appertaining to one species.

DECAPODA.   PENÆADÆ.
MACROURA.

GENUS PENÆUS. F<small>ABR.</small>

P<small>ALÆMON</small>.   Oliv.
P<small>ENÆUS</small>.   Fabr. Bosc. Latr. Leach, Edw.

Generic character.—*External antennæ* about as long as the body, the scale slightly decreasing and rounded at the apex, ciliated on the inner margin. *Internal antennæ* with the first articulation very broad, hollowed above, forming a cavity for the eyes; the outer margin armed with a tooth, and the inner furnished with a lamellar and ciliated appendage; the last two joints very short. The filaments of these antennæ are double and generally very short. *External pedipalps* long, slender, and pediform, furnished at their base with a long, curved, ciliated appendage. *Feet* all with a small appendage at the base; the *first three pairs* didactyle, increasing in length from the first to the third. *Carapace* with a prominent median crest, extending into a long toothed rostrum. *Eyes* very large and round. *Abdomen* large and much compressed; its posterior half carinated. *False feet* much enclosed by the lateral portions of the abdomen, terminating in two unequal ciliated plates.

One species only of this genus has been found on our coasts.

DECAPODA.
MACROURA.

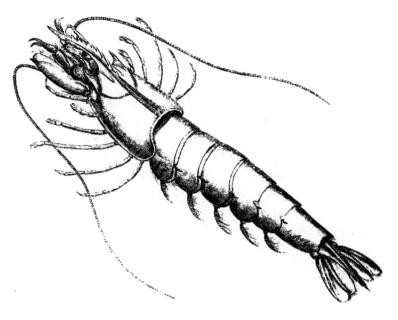

Penæus Caramote. Desmar.

Specific character.—Filaments of the internal antennæ shorter than the last two articulations. Thorax trisulcate posteriorly; rostrum bent downwards, above multidentate.

Alphæus caramote		Risso, Crust. de Nice, p. 20.
Penæus	,,	Desmar. Consider. sur les Crust. p. 225.—Risso, Hist. Nat. de l'Eur. Mer. V. p. 57.—Edw. Hist. des Crust. II. p. 413.
,,	*trisulcatus*	Leach, Mal. Brit. t. xlii.

The carapace is large and somewhat ventricose; the rostrum extending to the peduncle of the internal antennæ, armed above with numerous (about twelve) rather strong teeth, and beneath with one or two points only, the latter placed a little in front of the line of the eyes; on each side of the rostrum is a crest, which is continued backwards to near the margin of the carapace, thus forming a deep groove on each side of the median crest of the rostrum, the posterior portion of which is also traversed by a third longitudinal groove. A strong tooth at the anterior margin of the carapace, above the insertion of the internal antennæ, and at the outer and upper margin of the orbit, a smaller tooth at its inner side, and a third very small one at the posterior termination of a small lateral groove which passes backwards from the face of the second tooth. The eyes are round and remarkably large. The inferior or internal antennæ have two extremely short filaments, shorter than the last two articulations of the peduncle. The scale of the external antennæ is somewhat narrower and evenly rounded towards the extremity, and ciliated along its anterior and inner margin. The external pedipalps are pediform, gradually tapering to the extremity, and terminating in a small acute finger. The first three pairs of feet didactyle, increasing in length from the first to the third, which is rather the longest of the five. The hand of the first pair is the most robust, that of the second rather the longest: the fourth and fifth pairs are simple. The abdomen is much compressed, particularly posteriorly, and rises to an acute carina for the greater part of its length backwards. The fourth and fifth segments are notched on each side. The last segment, or medium portion of the tail, is long, narrow, triangular, acute, longitudinally grooved, furnished with a strong tooth on each side near the apex.

The length of the British specimen, figured by Leach, from the rostrum to the extremity of the tail is not more than three and a half inches. Risso gives five inches, and Milne Edwards seven as the length of those of the Mediterranean.

I have felt compelled, upon careful examination, to consider the British species of *Penæus*, to which Leach, probably from not having seen any Mediterranean specimen of *P. caramote*, gave the name of *P. trisulcatus*, as identical with that to which the former name has been given by Rondeletius, by Risso and Edwards. In this conclusion I am borne out by the opinion of the last-mentioned distinguished naturalist. Like some other of the Mediterranean species found on our shores, it is very rare with us. Leach mentions but two specimens known to him, nor is it mentioned in any of the local Faunas either of England or Ireland. The two specimens known were both taken on the Welsh coast.

GENERA CUMA, Edw., ALAUNA and BODOTRIA, Goodsir.

In the 13th volume of the " Annales des Sciences Naturelles," Dr. Milne Edwards described a small Crustacean under the name of *Cuma Audouinii*; but in his " Natural Hist. of Crustacea," he expresses his doubt whether this little animal be anything more than the larva of a decapodous form, and places it amongst other doubtful examples in an appendix.

In 1843, however, Mr. Harry Goodsir published in the "Edinburgh New Philosophical Journal," a very full and clear description of this and two other species of *Cuma*, and of two allied species which he considers as the types of two new genera, to which he gives the names respectively of *Alauna* and *Bodotria*. The whole of these I have ventured to consider *provisionally* as constituting a small family, probably belonging to the lower decapods, which appears also to be Mr. Goodsir's own opinion, though expressed with doubt, in which doubt I entirely agree. This author satisfactorily determined that they are perfectly developed animals and not mere larvæ.

As I have never had an opportunity of seeing the animals, I take the liberty of giving the whole of Mr. Goodsir's account of this remarkable family, which is too concise to require or admit of condensation.

For the anatomical details I refer to the plates illustrating the paper.

" During the summers of 1841 and 1842, I obtained a number of crustaceous animals, which I arranged promiscuously under the genus *Cuma* of M. Edwards, it being my intention to publish them at that time under this arrangement. I waited, however, until it could be satisfactorily proved whether they were perfect animals, or, according to the suspicions of M. Edwards, merely the larvæ of some Decapodous Crustacea. I have now satisfied myself that they are perfect animals, and at the same time have discovered the types of two new genera, which places the group in a still more interesting point of view.

" I have applied the name *Bodotria* to one of these genera, and *Alauna* to the other; the former being the ancient name of the Firth of Forth, at the mouth of which all these animals were got; and the latter, the ancient name of the river Forth.

" The latter of these genera (*Alauna*) may be the genus *Condylurus* of Latreille, as I have never seen that author's description; but whether it be so or not there cannot be any danger in applying the name *Alauna,* as *Condylurus* had been previously used amongst the Mammalia.

" As I had a greater number of specimens of the *Cuma Edwardsii* than of any of the others, I have been enabled to make out the structure of that species with greater minuteness.

" These animals are very like small prawns in their general appearance; but they bear perhaps in this respect a greater likeness to the species of the genus *Nebalia* than to any other known Crustaceans.

" The shell is hard and brittle, cracking under pressure. All the species are of a pale straw colour. The thoracic portion of the body is large and swollen; it is composed of six segments; the abdomen is longer; and is composed of seven segments.

" M. Edwards, in his Memoir on the genus *Cuma*, published in the 13th vol. of the Ann. des Sc. Nat., considers that the whole of the first and largest segment of the body constitutes the head. In all the specimens which I have dissected, I have found a suture running across this segment, immediately before the middle part of it; this is observed very distinctly in the *Cuma trispinosa*, in the *Bodotria arenosa*, and also in the genus *Alauna*. The first of these parts I consider to be the head; the second part as the first thoracic segment. To the first we find attached the rostrum, eyes, antennæ, organs of the mouth, and footjaws four in number. The second part bears the first pair of true ambulatory legs; these legs constituting (according to M. Edwards) the third pair of footjaws.

"The second thoracic segment is quite obsolete in M. Edwards's species (*Cuma Audouinii*); it is but slightly observed in the *C. Edwardsii*; in the *C. trispinosa*, however, it becomes quite apparent, being of considerable breadth at the dorsal portion. In the *Alauna rostrata*, also, we find this segment quite developed throughout its whole extent, and the second pair of thoracic legs arising from it.

"These two thoracic segments (the first and second) bear the compound legs in the genera *Cuma* and *Bodotria*, in which two genera the four following segments bear the four pairs of simple legs. In the genus *Alauna*, however, we find a different arrangement, there being an equal number of simple and compound legs, three pairs of each.

"The eyes in this tribe of animals are exceedingly small; they are pedunculated, but sessile,* and are placed very close together; they are situated near the posterior part of the head, a short distance behind the rostrum, and on the mesial line. They are covered by the shell, owing to which, and their proximity to one another, the animal is at first sight apt to be considered as monoculous. The rostrum is short and truncated in the genus *Cuma*; is almost altogether awanting in *Bodotria*, but is well developed in *Alauna*, being of considerable length and pointed.

"The antennæ undergo considerable changes in the different genera of this tribe. In *Cuma* we find the superior antennæ con-

* This passage appears to be inconsistent. The two great families of Malacostroca are essentially distinguished from each other by the eyes being relatively pedunculated (*Podophthalma*) or sessile (*Edriophthalma*).—T. B.

sisting of a single scale-like joint, armed with a number of strong spines; the inferior antennæ are five-jointed, being in general very little longer than the rostrum. In *Bodotria* the superior antennæ are altogether obsolete, and the inferior antennæ are very short. In *Alauna*, again, we find the antennæ more developed; the superior consisting of a single-jointed peduncle, and a long multiarticulate filament which is covered with hairs. The inferior pair are eight or nine-jointed, and are somewhat larger than the rostrum. The organs of the mouth consist of one pair of maxillæ, three pairs of mandibles, and two pairs of foot-jaws. These last organs will be found minutely described under *Cuma Edwardsii*, the species which I have been enabled to examine most minutely.

"The true legs may be classed into compound and simple. The compound legs, as we have already stated, are four in number in the genera *Cuma* and *Bodotria*; but six in *Alauna*. The first, or compound legs, are divided into two parts, the anterior or ambulatory, and the posterior or natatory. The simple legs are much shorter than the compound, and are more adapted for prehension; but they are unarmed with claws, and are seldom used for this purpose.

"The abdomen is moniliform, seven-jointed, in all the genera. The last joint is very small in the genera *Cuma* and *Bodotria*; but in *Alauna* we find this segment very much developed. All the genera have the sixth abdominal segment armed with a pair of long bifurcated styles. The genera *Cuma* and *Alauna* are quite free of appendages to the other abdominal segments; but in *Bodotria* we find that all the abdominal segments are armed with a pair of bifurcated appendages.

"Owing to the opacity of the shell, I have not been able as yet to make out the minute parts of the anatomy of these animals. The intestinal canal consists of a long straight tube, considerably dilated as it passes through the thoracic portion of the body; when it reaches the abdominal portion it suddenly becomes much narrower.

"The anal aperture is found in the seventh abdominal segment.

"The branchiæ are situated on each side of the thorax, immediately above the insertions of the legs, and approach, in their comb-like appearance, to those of the higher Crustacea. Interiorly, each of them is connected with the superior foot-jaws, and, excepting that connection, lies apparently quite free in a sac formed by the reflection of a thin transparent membrane, which lines the internal surface of the thorax. The superior part of the branchiæ consists of one continuous piece, which is bent in a hook-like manner at its posterior extremity; the branchiæ themselves arise from the inferior edge of this part, and are about sixteen or seventeen in number; they are not laminated like those of the higher Crustacea, but consist of one large piece, which is apparently composed of a great number of cells.

"The organs of generation are not apparent in the male, but in the female, and, especially when she is loaded with spawn, these organs are at once perceptible. They are very similar in their structure and appearance to the same parts in the female *Mysis*. They consist of four scales, which arise from the inferior edge of the thoracic segments. These scales are of an irregular oval shape, concave internally, and convex externally, and they are overlapped by one another. The eggs are of considerable size, and of a bright straw colour. It is from the genus *Cuma* only that these observations were taken in regard to the organs of generation.

"When a portion of the skin, or shell rather, is placed under the microscope, it presents a very beautiful appearance; it apparently consists of a great number of nuclei, arranged in some degree of order. These nuclei are stellated, and here and there larger nuclei may be observed, the edges of which are quite smooth.

"The structure of these animals is so peculiar, as to render the assignation (at present) of a proper place in a natural arrangement of the class, a point of very considerable difficulty. This arises in a great measure, without doubt, from our very limited knowledge of the class. I rather think, however, that they should be ranged among the lower *Decapoda macroura*.

Genus CUMA (*Edwards*).

Generic Characters.—The superior antennæ are single-jointed, and scale-like; the inferior antennæ are five-jointed. The caudal styles have the double terminal scales biarticulate, the last of which is always the shortest.

Cuma Edwardsii, mihi.

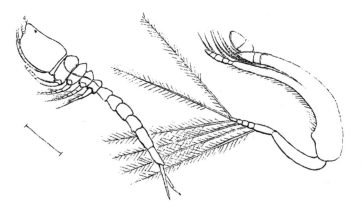

C.—With the superior antennæ rhomboidal; with the ambulatory division of the first pair of legs, with the first joint bent at an obtuse angle; with the thumb-like process single-jointed, and with the last joints clavate. Length 4 lines. Hab. Frith of Forth.

Description.—The whole animal is of a fine straw-colour, with a delicate tinge of pink, which is brighter in certain lights; the shell is quite rough, which is caused by the great number of shallow foveæ with which the whole surface is thickly covered. This, and the following species, are perhaps the smallest of the genus; at the same time, they are much thicker and stronger in proportion to their size than the other species. The rostrum is short, thick, and suddenly truncated obliquely. The antennæ are minute; the first or superior pair are almost obsolete; they consist of one joint only, which is rhomboidal: the extremity of each is armed with several strong but minute hairs or spines; they arise from the truncated extremity of the rostrum. The inferior antennæ arise from the inferior surface and base of the rostrum; they are considerably larger than the superior pair; they are five-jointed, the third joint being the longest, the fifth or last is extremely small, and is armed with three very strong pointed and articulated spines. These pair of antennæ are somewhat longer than the rostrum. The footjaws are rather powerful, and have a great resemblance to the following pairs of feet. The first, or superior pair, are the smallest; the first joint is of considerable length, being equal to all the others combined; it is rather bent and broad, and is armed at its distal extremity with two thumb-like processes or tubercles. Two very long and slender spines, which are almost as long as the foot-jaw itself, arise from the middle part of this segment; the external spine is free of spinules altogether, but the internal is armed, on its external edge only, with a great number of articulated spinules. The second segment of this footjaw is very short, and its posterior edge bears two very short articulated spines of equal length; these spines are spiniferous. The third segment is almost equal in length to the first, and, like the second, also gives rise to nine or ten articulated and spiniferous spines. The fourth segment is small and rounded, being also armed on its posterior edge with simple spines. The fifth segment is thumb-like, and spinous on its posterior edge.

The external pair of footjaws are much larger than the internal; they are five-jointed, and are armed in the same way as the first pair, except that the external edge of the first segment is armed at regular intervals with small tufts of very fine hairs; the extremity of the second segment is also armed with a very long articulated and spiniferous spine. These two extremities just described are in general lying in such a way as to cover the organs of the mouth.

The first two pairs of legs are constantly concealed beneath the carapace when the animal is at rest, covering the footjaws and the organs of the mouth, and appear only to be used when the animal is swimming. The anterior or ambulatory division is five-jointed; the first joint is about twice the length of all the others combined; it is considerably bent and very broad; its internal edge is armed at regular intervals with pennicillated tufts of hair; the three following segments are quite free of spines, but the last is armed at its extremity with a strong claw and two smaller spines. An articulated thumb-like and chelate joint arises from the extremity of the first segment, immediately internal to the last four segments. The natatory or posterior division of this leg is multiarticulate; the first two segments are longest, being equal in length to the first segment of the anterior division; the remaining segments are minute, about nine or ten in number, each of which gives off a very long spiniferous setum, which is articulated at its distal half. The second thoracic leg of this species presents to us one of those beautiful and delicate structures which it is impossible either to describe or to delineate with even a remote degree of accuracy. The ambulatory division is very long and slender, six-jointed; the first joint is long and very much flattened, but tapers from the middle towards its distal extremity, which is armed with a very long and pointed spine; the following joints are all equal to one another in length, except the last, which is minute. The natatory division of this leg is seven- or eight-jointed, and is equal in length to the first segment of the other division. The last five segments are all armed with long articulated and spiniferous setæ, which smaller spines are again spinulose. The four following pairs of legs are simple, that is, they are merely ambulatory; they are all six-jointed, and are very spiny. The segments of the body from which they arise are all ovoid, their dorsal edge being sharp and pointed.

The abdominal portion of the body is long and slender, seven-jointed and moniliform; the last joint is minute, and lies between the caudal styles which arise from the extremity of the sixth segment; these styles are of no great length in this species; they are composed of three parts; each style consists of a long-jointed peduncle, from the distal extremity of which two biarticulated scales arise; these scales lie one above the other. The first segment of the peduncle is somewhat longer than the sixth abdominal segment; the first segments of the scales are about half the length, and the last segment about one-fourth the length of the peduncle; the inner edge of the superior scales is armed with a number of long, pointed, and articulated spines. The spines which arise from the inner edge of the inferior scales are more numerous; they are all bent, their points being turned backwards; the convex or anterior edges of all these spines are very much serrated.

I have named this species after M. Edwards, the founder of the genus, and the leading crustaceologist of the day.

Cuma Audouinii. Edwards.

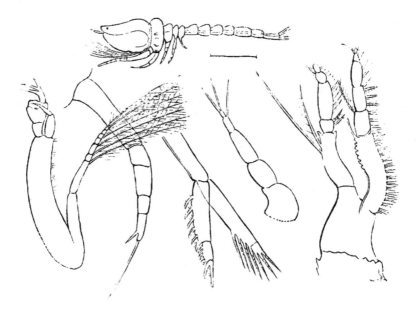

C.—With the superior antennæ very small; with the first joint of the ambulatory division of the first pair of legs almost bent at right angles; the terminal joints oval, and the thumb-like process multiarticulate. Long, four lines to five. Hab. Frith of Forth.

Description.—Under casual observation this species is very apt to be mistaken for that last described, but by careful examination the difference is found to be very material. In its general appearance, this species resembles the *Cuma Edwardsii*. The first thoracic segment, however, is longer and not so rounded; the rostrum is shorter and more pointed, and the eyes are larger; the flattened surface on the sides of this species is not so decided. The second thoracic segment is more hid; the third is larger, ovoid, and rounded; the adjoined scale projects backwards; the fourth segment is of the same shape as the third, but not nearly so large; the fifth ends in a sharp point, both superiorly and inferiorly; the sixth thoracic segment is clavate. The superior antennæ are very small, and scarcely to be distinguished from the rostrum. The inferior antennæ are very similar to those of the *Cuma Edwardsii*. The footjaws are also similar in their structure to those of the last-described species; the ambulatory division of the first leg is five-jointed; the first joint is very much bent, and is of considerable breadth; the two last joints are quite oval, and the last nonchelate. The internal thumb-like process, instead of being composed of one joint only, as in the last described species, consists of four or five segments, which are all armed with short spiniferous and pointed spines; the natatory portion of this leg is multiarticulate, the extreme joints being very small, so as to place the long spiniferous setæ very close to one another.

The second pair of legs are very short. The last four pairs of legs are similar in their structure to those of the last described species. The abdomen and caudal fins also bearing a similar resemblance.

This species is apparently the *Cuma Audouinii* of M. Edwards, but whether it is or not I cannot be quite certain.

Cuma trispinosa, mihi.

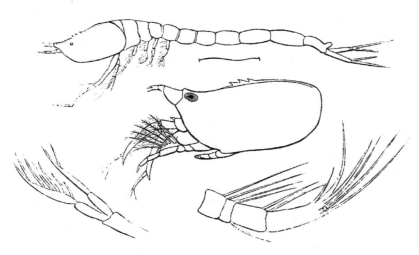

C.—With the dorsal ridge of the carapace surmounted by three spines, with the ambulatory division of the first pair of legs extremely short, and with the second thoracic segment well developed. Long, eight lines. Hab. Frith of Forth.

Description.—This is a most characteristic species, and brings out several points of material consequence in the character of the genus. This species has the body quite smooth, and of the same colour as the preceding. It is the largest of all the species, but is more slender. The thoracic segments are not so deep as those of the preceding species, and the lateral compression is awanting. The rostrum is sharp-pointed, and bent considerably upwards; the eyes are small, and the dorsal ridge immediately behind the eye is surmounted with three thick short spines. The second thoracic segment is of considerable extent at its dorsal part, but is quite obsolete at the middle; it again, however, makes its appearance at its inferior part, where it supports the second pair of compound legs. The four following segments gradually decrease in size:—the superior antennæ are of considerable size, oblong and spinous. The inferior antennæ are much longer than the rostrum. The ambulatory division of the first pair of legs is extremely short, and the first joint is of no great breadth. The natatory division it about the same length as the first joint of the anterior division.

The second pair of legs are very long and slender; the first segment is not broader than the following joints, and is armed internally at its extremity with a very long spine.

The simple feet are extremely spiny.

The abdominal portion of the body is very long and slender, the fifth segment being the longest. The caudal styles are long, slender, and pointed; the internal scale has the last joint pointed and armed with two spines; the last segment of the external scale is more obtuse.

Genus ALAUNA, mihi.

Generic Characters.—The superior antennæ are composed of a peduncle and a multiarticulate filament. The inferior antennæ are eight-jointed. The first three pair of legs are compound. The internal scale of the caudal style is composed of three segments, and the external of one.

Alauna rostrata, mihi.

Description.—The whole animal is of a beautiful bright straw colour, inclining to yellow. The thoracic portion of the body is very large and swollen. The first segment or carapace is almost oval. The rostrum is long, pointed, and is bent upwards at its extremity. The eyes, which are of considerable size, are situated at the base of the rostrum. The superior pair of antennæ are very slender, consisting of a delicate filament covered with hairs, which arises from a short peduncle; these antennæ are almost equal in length to the rostrum.

The inferior antennæ are much longer, consisting of eight joints slightly spinous; the distal extremity of the third is armed with a strong multiarticulate spine. The footjaws are seen projecting considerably beyond the edge of the carapace; they are very spiny, and the last joint but one is armed with a long articulated spiniferous spine.

The first pair of legs are extremely short; the thumb-like process at the extremity of the ambulatory division is single-jointed and spiniferous. The second pair of legs are also short. The ambulatory division of the third pair of legs is very long and slender, being almost as long as that of the second pair of legs; the fifth joint is the longest. The natatory division is as long as the first four joints of the ambulatory. The simple legs are very spiny on their anterior edges.

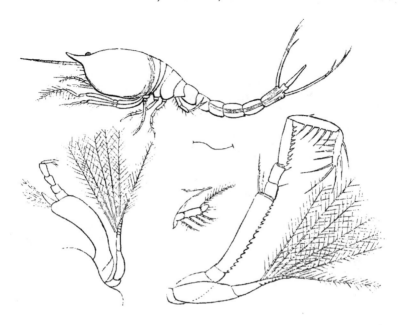

The abdomen is short and thick, seven-jointed, the last joint being produced into a long spine which is spiniferous on either edge; the anal aperture is seen near the base of this segment. The caudal styles arise from the sixth segment, and they are much more complicated that those of the foregoing genera. The first segment is slightly clavate, longer than the seventh abdominal segment, and armed with a single row of spines on its inner edge. The internal scale consists of one joint only; it is very spiny, and is about half the length of the external. The external scale is composed of three joints, the first two of which are equal in length to one another; the third is about twice the length of both of these, and is very spiny at its extremity. Long, half-an-inch. Hab. Frith of Forth.

Having only obtained one specimen of *Alauna rostrata*, and one also of *Bodotria arenosa*, I have not been able to examine the structure of these two genera satisfactorily.

Genus Bodotria, mihi.

Generic Characters.—The first, second, third, fourth, and fifth abdominal segments are each armed with a pair of bifurcated finlets. The two terminal scales of the caudal styles are single-jointed.

Bodotria arenosa, mihi.

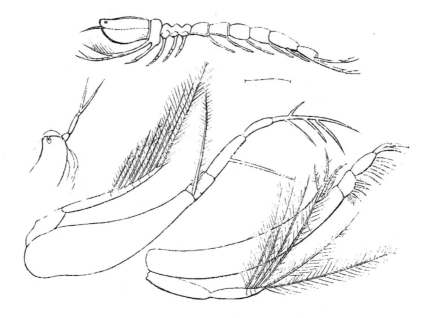

Description.—The carapace is almost oval, rostrum awanting, that part of the carapace being merely rounded off. The superior antennæ are quite obsolete. The inferior pair are of considerable length, and are terminated by means of two long spines.

The ambulatory division of the first pair of legs has the first joint of a very great size, being very much flattened and slightly curved. The four remaining joints, together with the internal thumb, are very spiny. The natatory division of the leg is six-jointed, the four last joints giving rise to as many long spiniferous spines, which are articulated at their distal halves. The external edge of these spines are spiniferous at the articulated half only. The ambulatory division of the second pair of legs has the first segment very broad, and tapering gradually towards its distal extremity, from which arises a very long, articulated, and spiniferous spine.

The abdominal finlets are five in number. They are composed of two parts, viz., the first or pedicle, and the second or bifurcation; the pedicle is of considerable length, from the extremity of which there arise two scales, which are armed on their margins with long spiniferous spines, which are much longer than the finlet itself.

The first segment of the caudal styles tapers very slightly, and the two terminal scales are each of them single-jointed, and end by means of very fine points. The external is armed at its extremity with two spines. Long, five lines.

This genus forms doubtless a link between the *Stomopoda* of M. Edwards and the higher Crustacea.

"In their habits all these animals seem to agree. I have not been able to observe anything peculiar in them. They swim with very great rapidity, and on stopping they fall to the bottom on the sand or gravel, without attempting to lay hold of anything, as I have already remarked, seldom using their feet as a means of prehension. They free themselves with great dexterity from any weight which may happen to fall on them. I have often placed the point of a needle on their thorax and pressed them down into the sand; the animal immediately frees itself with very little apparent trouble, by means of its tail. The extremity of the tail is placed against the needle with one of the styles on either side of it, and by pressing upwards in this way, it soon regains its liberty.

"They frequent sandy banks, and chiefly those where there is a little sea-weed."

GENUS MYSIS, Latr.

Cancer.	Muller, Otho, Fabr.
Mysis.	Latr. Lam. Leach, Edw.

Generic character.—*External antennæ* inserted beneath the internal, the first joint giving attachment to a laminar appendage, similar to that in the *Palæmonidæ*, which is much elongated and ciliated on the inner margin; the two succeeding joints of the peduncle slender and cylindrical, the terminal filament filiform, and longer than the *internal antennæ*, which are inserted beneath the eye, near the median line, and have two terminal filaments. *Pedipalps* consisting of two pairs entirely pediform. *The first pair* short, composed of three distinct branches; the internal portion pediform, of five joints, hairy, and doubled upon itself in front of the mouth; the middle branch or palp elongated, and composed of numerous articulations; the basilar joint very large, with a ciliated strap-shaped process on each side; the third or external branch, or flabelliform appendage, is represented by a semimembranous scale directed upwards, and lying under the margin of the carapace. *Second pair of pedipalps* of the same form, but wanting the flabelliform appendage. *Feet* of six pairs, composed of corresponding elements with the external pedipalps and five pairs of feet in the Decapoda; each consisting of two branches, decreasing in length from before backwards, and formed for swimming; the first four pairs have no flabelliform appendage; the last two are furnished with it. This part in the male is very small, but in the female it is greatly developed, and forms on each side a broad plate bent under the sternum, the two thus forming a pouch, in which the eggs are first deposited, and within which the young are secluded, and pass the early period of their life. *Carapace* covering only the anterior part of the thorax, the two sides bent downwards and inwards so as to be applied against the base of the feet; anteriorly it becomes very narrow, and terminates in a short flattened rostrum. *Eyes* large, short, with the base hidden under the anterior margin of the carapace. *Abdomen* very slender, tapering, elongated, nearly cylindrical. *Tail* as in the macrourous DECAPODA.

No distinct branchial apparatus has as yet been observed in this remarkable genus; and, as is observed by Dr. Milne Edwards, "The only appendage which appears to be so modified in its structure, as to become more adapted than the rest of the body to serve the purposes of a respiratory organ, is the *lash* of the first pair of pedipalps, which in other respects are similar to those found in numerous species possessed of branchiæ." It is, however, not at all improbable that this may be the true organ of respiration.

The development of the young in this genus, as well as their anatomy generally, has engaged the attention of the late Mr. J. Vaughan Thompson, and a very elaborate monograph of their structure will be found in his "Zoological Researches," to which the reader is referred for full information.

The affinities of the family *Mysidæ* are very incorrectly indicated by the position which Dr. Milne Edwards has assigned them amongst the *Stomopoda*. In almost all the essential points of structure they are certainly more nearly allied to some of the *Decapoda*; but as they are also remote even from these, I have not considered it right to reduce them to that group, or to attempt to fix their natural relation to the two groups, particularly as a local Fauna does not offer the best vehicle for changes in general arrangement.

MYSIDÆ.

STOMOPODA.

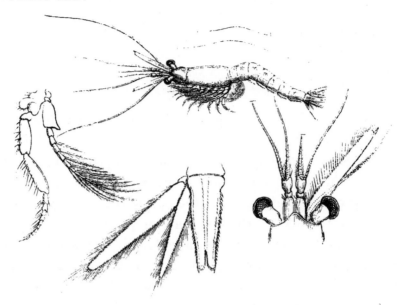

Mysis chamæleon. J. V. Thomps.

Specific Character.—Middle plate of the tail bifurcate; rostrum obtuse, not more than one-third the length of the ocular peduncle.

Mysis spinulosus?	LEACH, Trans. Lin. Soc. XI. p. 350.—DESMAR. Considér. sur les Crust.—EDW. Hist. des Crust. III. p. 457.
,, *Leachii?*	J. V. THOMPSON, Zool. Researches, p. 27.
,, *Chamæleon,*	Ib. p. 28, t. ii. fig. 1—10.—EDW. l. c. p. 457.

THE general form of this species is much elongated. The carapace slender, terminating in a very short rostrum, in some scarcely projecting, in others forming an obtuse triangle, and never extending more than one-third the length of the ocular peduncle. The internal antennæ have the peduncle somewhat club-shaped, the first joint being elongated, cylindrical, and small, the last two, and particularly the terminal one, much broader, and both very short; the whole peduncle does not extend much more than one-third the length of the scale of the external antennæ: the scale becomes a little narrowed forwards, is obliquely truncate at the apex, with a small tooth on the outer angle, and it is ciliated with rather long hairs on the inner side and at its extremity. The middle plate of the tail is bifurcate at the apex, longitudinally grooved on each side of the median line, minutely toothed on the sides, and with a stronger tooth on each apex. The lateral laminæ are ciliated on all sides with long hairs; the inner is long-lanceolate, and acute; the outer is longer, and rounded at the extremity.

It often reaches the length of an inch and a quarter.

"Nothing," says Mr. Vaughan Thompson, "can shew the fallacy of colour in distinguishing the species, more clearly than the variety of tints which *Mysis chamæleon* assumes, as it occurs here in the river Lee and the harbour of Cove, and which have suggested its trivial name; in the upper part of the river, below the city of Cork, it occurs of different shades of grey, inclining at times to black, having invariably the greater part of the anterior scales, inner branch of the inferior antennæ and joints of the outer laminæ of the tail, black, and the fringe of the scales tinged with pink; lower down amongst the littoral fuci, it takes various tints of brown; and those obtained from sites abounding in Zostera and Ulvæ, present us with green colours of greater or less intensity."

I have quoted the above account of the variation of colour in this species in the author's words, in order to shew that difference of colour alone can afford no ground for considering this species as distinct from the *spinulosus* of Leach. And yet Mr. Vaughan Thompson, in his description of the latter (which he names *M. Leachii*), gives colour as the only tangible distinction. I am decidedly of opinion that they constitute but one species, and I have retained the name of *chamæleon*, as *spinulosus* is equally applicable to various other species of the genus.

This is, perhaps, the most common and the most widely distributed of our native species. I have received it from various parts of the coast, both of England and Ireland, but from no place in such numbers as from Weymouth, where it sometimes swarms. My late lamented friend, Mr. William Thompson, informed me that he has taken this species in numbers from the stomach of Corregonus Pollan, caught in Lough Neagh, shewing that it occasionally inhabits fresh water.

STOMOPODA. MYSIDÆ.

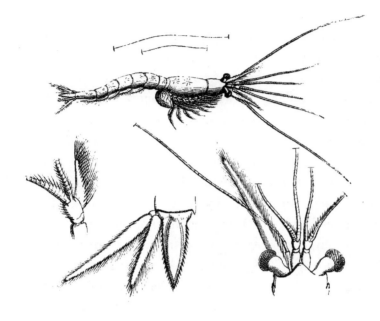

Mysis vulgaris. J. V. Thompson.

Specific Character.—Middle plate of the tail lanceolate, the apex entire; rostrum very short, obtusely triangular, extending to about half the length of the ocular peduncle; antennal scale nearly as long as the carapace.

Mysis vulgaris, J. V. THOMPSON, Zool. Researches, p. 30, t. i.—EDW. Hist. des Crust. III. p. 459.

GENERAL form less robust than in *M. chamæleon.* The carapace slender, somewhat cylindrical, slightly constricted at its anterior third, terminating in a very short, obtusely triangular rostrum, which scarcely extends to the middle of the ocular peduncle. The peduncle of the internal antennæ much resembling that in the former species; scale of the external antennæ not less than four times the length of the peduncle of the internal; subulate, obtuse at the points, ciliated on both sides, and without an apical tooth. Middle plate of the tail lanceolate, acute, spinulose on the sides; lateral plates somewhat subulate, ciliated on each side.

Mr. J. V. Thompson appears to have been the first to distinguish this species of *Mysis*, to which he gave the name *vulgaris*, probably from its being the most common species in the locality where he found it. It is, however, more rare on our coasts, and probably, as it had escaped detection, also on those of other countries. The specific name which he assigned to it is always one of doubtful propriety, and I would fain have assigned to it that of the distinguished naturalist by whom it was first discovered and described, but from a disinclination ever to change a specific name excepting under urgent circumstances.

It appears to be a local species, and as far as we are at present able to say, is principally found on the Irish coast; "abounding in the Lee," says Mr. J. V. Thompson, "even up to Cork," and I have specimens collected by Mr. W. Thompson in Belfast Lough. From the former species it may be at once distinguished by several prominent characters, particularly the longer antennal scale, and the simple acute apex of the middle caudal lamina. Its colour is pale, translucent grey.

The following is Mr. Vaughan Thompson's account of its habits. "They swim with the body in a horizontal position, and abound in the Lee, even up to Cork, from the early part of spring to the approach of winter; during the still period of the tide at low water, they repose upon the mud and stones at the bottom of the river, and as the tide rises, may be observed forming a wide belt, just within its margin, the youngest swimming nearest to the shore, the oldest further out, and in deeper water: they appear to be mostly females, the males being few in proportion. This species contributes towards the food of various young fish, from which they frequently escape, by springing up out of the water."

Is it not probable that *M. integer* of Leach (*Scoticus* of Thompson), may be identical with this species? His description is too imperfect to enable us to ascertain this, but they agree in such particulars as are known, excepting in size and colour, both of which are variable characters.

The following is his account of the species.*

"Tail with the middle lamella entire; length one-third of an inch. Colour pellucid cinereous, spotted with black and reddish brown. Eyes black. Females more abundant than males.

"At low tide, near Loch Ranza [in the Isle of Arran], the pools were full of this species, swimming with its head uppermost, and its eyes spread, making a most grotesque appearance."

* Lin. Trans. XI. p. 350.

STOMOPODA.   MYSIDÆ.

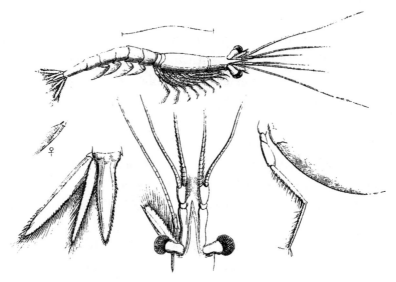

Mysis Griffithsiæ.   Mihi.

Specific Character.—Middle plate of the tail lanceolate, constricted near the base, apex entire, slightly obtuse; rostrum lanceolate, extending beyond the penultimate joint of the peduncle of the internal antennæ; antennal scale scarcely longer than the rostrum.

Mysis rostratus ?   Guer. Iconogr. Crust. t. xxiii. f. 2.

Carapace much elongated, and more slender than in either of the former species, terminating forwards in a long, acute, lanceolate rostrum, extending beyond the penultimate joint of the peduncle of the internal antennæ, which is itself longer and more slender than in the former species. Scale of the external antennæ shorter than the peduncle of the internal, rounded at the apex, with a small external spine, ciliated only on the inner margin. Abdomen very slender and tapering; middle plate of the tail lanceolate, somewhat constricted near the base, the apex entire, very slightly obtuse, the margins spinulose; the inner lateral lamella very narrow, tapering regularly to the end, fringed on both sides with long hairs; outer lamella broader and longer, fringed with long hairs on the inner margin, and at the apex, and with a few short stiff hairs only on the anterior third of the outer margin.

Length three quarters of an inch.

I can scarcely persuade myself that this can be the species figured, but not described, by Guerin under the name of *M. rostratus*, although the characters in many respects agree with his figures. The form of the rostrum, of the antennal scale, of the peduncles, of the eyes, of the tail, and indeed of every part figured, although bearing a general resemblance, differs so much in detail that we are left in the dilemma either of considering the representations worthless from their inaccuracy, or of giving a distinct specific name to ours. As no description exists of M. Guerin's species, I have adopted the latter alternative, and have named it in honour of a lady to whom natural history is greatly indebted, and from whom I received the only specimens of this species known. Mrs. Griffiths obtained them at Torquay.

GENUS THYSANOPODA, Edw.

Generic character.*—*External antennæ*, as in *Mysis*, inserted beneath the internal, and furnished with a small antennal scale, the basal joint broad and almost globular. *Internal antennæ* inserted close beneath the eyes; furnished with two filaments. *Pedipalps* two pairs, entirely pediform, perfectly resembling the legs themselves. *Feet* similar to each other and to the pedipalps, excepting the last pair; the basal joint short and thick; the stalk very long, and furnished with long hairs on the inner side; the palp short, lamellar, and hairy; the last pair of legs much shorter, consisting only of the palp, which is more developed. *Carapace* as in *Mysis*. *Abdomen* with the lateral processes more developed than in that genus; the first five segments furnished with natatory false feet—the appendages of the sixth segment forming the lateral caudal laminæ, and the seventh constituting the central, slender, and furnished at the apex with two needle-shaped appendages. *Branchiæ* external, consisting of eight pairs, attached at the base of the several pairs of thoracic natatory members; the pedipalps and true feet increasing in size and development from the first to the last. They consist of a stem, or stalk, each furnished with numerous lateral branches. *Eggs* contained in a pair of oval sacs dependent from within the base of the posterior feet.

This very remarkable genus of Mysidæ was first described by Dr. Milne Edwards, in the Annales des Sciences, from an Atlantic species which he found in the collection of crustacea formed by Mons. Reynaud, and placed in the

---

* In the engraving at the head of the description of our species, are the details of all the essential parts; for the beautiful and accurate delineation of which I feel greatly indebted to Mr. Westwood's well known care and accuracy.

Paris Museum. The remarkable peculiarity in the respiratory apparatus distinguishes it at once from every other form of crustacea, and notably from its congeners in the same family; and the situation of the ova, which I am enabled to supply from the new species about to be described, is not less remarkable.

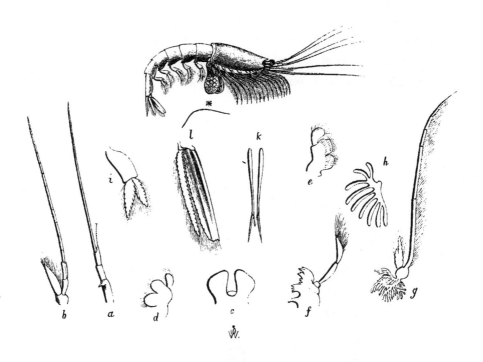

Thysanopoda Couchii.

Specific Character.—Branchiæ with only one series of leaflets. Middle point of the trifid apex of the central caudal lamina, not half the length of the lateral ones.

THE general aspect of this curious species indicates at once its near relation to Mysis, which the detail of its organization confirms. The present species differs from *Th. tricuspida*, the species on which Dr. Milne Edwards founded the genus, in several points,—the most striking of which are the following. In Edwards's species the branchiæ, in addition to the primary leaflets ranged in a single series along the stalk, have very numerous secondary filaments attached to them—a circumstance which I have not observed to exist in a single case of the numbers I have examined of our present species. Judging from the figures in Dr. Edwards's plates, the carapace in the present species is smaller and more cylindrical; the cleft in the lower lip is more hollowed, the palp of the thoracic feet is less developed, the abdominal false feet are shorter, and very differently formed; the middle lamina of the tail also presents some difference in the relative length of the middle and lateral points of its tricuspid apex. One of the most interesting circumstances in the organization of this species is the form of the ovisacs, which, instead of being mere pouches closely adherent to the thorax, are dependent from their attachment by a distinct peduncle. This structure was unknown until I detected it in a single individual, the only female amongst a large number of specimens sent to me by my friend Mr. Couch, who obtained them on the Cornish coast, from the stomach of a mackerel, which appeared to have been making a feast of this rare and interesting little crustacean. The following account has been kindly furnished to me by that gentleman, and shews that it can scarcely be considered as an ordinary inhabitant of our coasts. " The mackerel from which the curious shrimps Thysanopoda were taken, were caught almost at mid-channel, or almost ten leagues from us; perhaps seven or eight south of the Lizard; and I have not seen any since, although I am much in the habit of searching the stomachs of mackerel and other fishes. There were myriads in the stomachs of the mackerel at the time when I obtained those which I sent you." I have dedicated the species to that indefatigable and acute observer, to whom we are indebted for so many valuable contributions to natural science.

The following is a description of the details of the woodcut; *a*, superior antennæ; *b*, inferior antennæ; *c*, lower lip; *d*, first maxilla; *e*, second maxilla; *f*, mandible; *g*, one of the thoracic feet, with branchia attached; *h*, a branchia; *i*, abdominal false foot; *k*, middle caudal lamina; *l*, lateral caudal laminæ.

GENUS SQUILLA.

Squilla.	Rondel.
Cancer.	Lin. Herbet.
Squilla.	Fabr., Latr., Leach, Desmar. Roux, Edw., &c.

Generic character.— *Antennary segment* moveable, nearly quadrilateral. *External* or *inferior antennæ* inserted on each side the antennary segment, beneath the anterior margin of the carapace; the first and second joints of the peduncle short and thick, the latter bearing at its extremity a broad, oval scale; the terminal filaments slender and short. The *internal* or *superior antennæ* attached to the anterior margin of the antennary segment, and composed of a tri-articulate peduncle terminating in three filaments of moderate length. *Mouth* situated under the posterior third of the carapace. First pair of *thoracic members* forming a pair of robust claws, of which the terminal joint is furnished with long and sharp teeth, and is capable of being doubled upon the penultimate joint, into a groove of which it is received, forming a powerful prehensile implement. The three pairs following the claws are small, and terminated by a rounded hand, with a single finger, forming a single claw like that in *crangon;* the three posterior thoracic members much smaller, slender, cylindrical, and furnished with a styliform appendage attached to the extremity of the antepenultimate joint. *Carapace* longer than broad, divided by longitudinal sulci into three portions; not covering the first two cephalic nor the last four thoracic segments. *Abdomen* rounded above, each segment furnished with a pair of broad natatory false-feet, the basilary joint quadrilateral, each bearing two lamellar branches, the exterior of which gives attachment on its posterior face, and close to the peduncle, to a tufted *branchia.* The last segment of the abdomen is very large, forming the middle plate of the tail, the lateral portions of which are formed as usual by the appendages of the sixth segment; the basilary joint of these is very robust, and is prolonged into a long pointed scale, which stretches out beneath and between the two terminal branches.

It has been observed by Dr. Milne Edwards, that the distinctness and separation of the normal segments, especially those of the head and thorax, are carried further in this family than in any other form amongst the crustacea; and in this view it may be considered as offering the nearest approach to the typical structure, and the key to the homologies of the class. Some further allusion to this circumstance will be found in the introduction, and it will be sufficient here to refer the reader to the descriptions given by the excellent author just named, of the characters of the family, and of the different genera; and, in connexion therewith, to the plates in which the details of the external anatomy are given.* The species of the family are very widely distributed; and even of the genus Squilla, the coasts of Europe, Asia, Africa, and America, furnish examples. Of the three species which are known to inhabit the Mediterranean, two have now been found upon our South-western coast, both first discovered by the acute and indefatigable researches of Mr. Couch.

In the characters given above I have included those which are most characteristic of the family of Squilladæ, as well as those which are distinctive of the genus, as this is the only generic representation of the family indigenous to Britain.

* Hist. Nat. des Crust. II. p. 509 et seq. pl. 1, &c.

STOMOPODA. SQUILLADÆ.

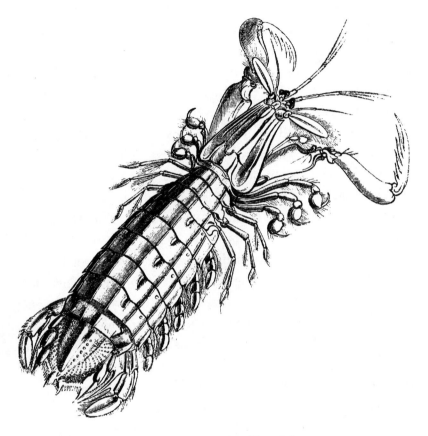

Squilla mantis, Rondel.

Specific Character.—Prehensile finger with six long teeth; abdomen with eight longitudinal crests, the two central ones near together; posterior margin of the middle portion of the carapace straight.

Squilla mantis, RONDEL. Poissons II. p. 397.
Cancer (mantis) digitalis, HERBIT. II. t. xxxiii. f. i. p. 92.
Squilla mantis, LATR. Hist. des Crust., &c., VI. t. lv. f. 3, p. 278.—
RISSO, Hist. des Crust. des Env. de Nice, p. 113.—
Hist. Nat. de l'Eur. Mérid. V. p. 85.—EDW. Hist. Nat. des Crust. II. p. 52.

THE carapace of this species is much narrowed anteriorly, and the anterior angles are slightly spiniform; the rostral plate semiovate; the middle portion of the carapace has a longitudinal median crest, which bifurcates anteriorly, and it is separated from the lateral pieces by a deep groove, which is continued transversely to separate the posterior portion; the lateral pieces have on each side two raised lines or crests, the outermost of which extends back to near the posterior margin. The claws very long and robust; the last joint furnished with six sharp, slightly curved teeth, inclusive of the extremity; the next joint with a deep groove for the reception of the last when closed, and the inferior margin of the groove is denticulated, and furnished with three moveable teeth at the base. The three posterior segments of the thorax with four crests. The abdomen is very broad and thick, broader and flatter towards the extremity, and having eight distinct crests, including the lateral margin of the segments; the two middle ones are nearer to each other than the others, and on the sixth segment terminate each in a sharp spine. The last abdominal segment (middle lobe of the tail) about as long as it is broad, furnished with a high median crest, terminating in a tubercle, at a short distance from the margin—the surface generally marked with a number of impressed points, arranged in curved lines. The margin is raised and thickened, and furnished with two pairs of lateral spines or tubercles, and there are two pairs of sharp strong spines on the anterior portion, with numerous small denticuli between them. The laminar prolongation of the basilary joint of the appendages to the sixth segment extends backwards as far as the external caudal scale, and is furnished with two very strong pointed horns. The first joint of the outer caudal lamina is strong and thick, and furnished, on its outer margin, with several strong spines which appear to be moveable.

The length of the English specimen, which is a female, from the frontal plate to the end of the tail, is four inches and a half. I have a male specimen from the Mediterranean which is no less than six inches.

Mr. Couch, to whom I am indebted for the specimen above referred to, informs me that "the Squillæ were brought from the distance of about a couple of leagues, where the bottom is rocky, with some spots of sand."

STOMOPODA. *SQUILLADÆ.*

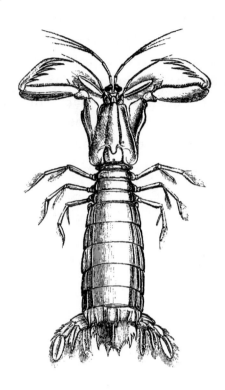

Squilla Desmarestii, Risso.

Specific Character.—Prehensile finger with five teeth; abdomen with four longitudinal crests, the middle portion smooth, excepting the sixth segment, which has two additional elevations.

Squilla Desmarestii, Risso, Crust. de Nice, t. ii. fig. 8, p. 114.—Hist. Nat. de l'Eur. Mérid. V. p. 86.—Desm. Consid. sur les Crust. p. 251.—Latr. Encycl. X. p. 471.— Roux, Crust. de la Médit. t. xl.— Edw. Hist. Nat. des Crust. II. p. 523.—Couch, Cornish Fauna, p. 81.— Yarrell, Loud. Mag. VI. p. 230.

The present species of Squilla differs in many striking characters from the former. The carapace has nearly the same general form, but is comparatively rather less narrowed anteriorly; it is less strongly marked, and the grooves and elevations neither so numerous nor so distinct. There is scarcely any trace of longitudinal crests, and it is wholly without spines on the anterior portion. The falciform finger of the claws is armed with five sharp teeth; the penultimate joint has the upper margin of the groove most minutely denticulated. The four exposed thoracic segments are smooth. The abdomen has on each side two low longitudinal crests, and the sixth segment two additional ones near the centre; the remainder of the surface is smooth. The terminal segment has a median crest, and the margin is furnished with six strong teeth, the interspaces being minutely denticulated. The laminar prolongation of the basilary joint of the appendages of the sixth segment projects nearly in the same proportion as in *Sq. mantis;* and the lateral caudal scales do not offer any striking peculiarity.

The length of full-sized specimens is three inches and a quarter.

This remarkably pretty species was first distinguished by Risso, who gives a description and an indifferent figure of it in his "Crustacés des environs de Nice," and repeats the description in his subsequent work, " L'Histoire Naturelle de l'Europe méridionale." A beautiful figure is given by Roux in his unfortunately unfinished work on the Crustacea of the Mediterranean. Its first occurrence on our coasts is recorded by my valued friend Mr. Yarrell in the sixth volume of Loudon's Magazine, with a figure, which is, however, very fallacious, from its being taken from a specimen which had become corrugated in drying. This specimen, with another, was taken by Mr. Couch on the coast of Cornwall, where they were found amongst seaweed; and thus another interesting addition is made to those which I have already had to record, to the partial identity of the Mediterranean Marine Fauna with that of our southern coast. The same fact is recorded by Mr. Couch in his Cornish Fauna.

I am lately informed by Mr. A. G. More of Bembridge, Isle of Wight, and of Trinity College, Cambridge, that it has also been taken repeatedly off Bembridge, by the fishermen of that place, on a muddy bottom grown over with "grass" (*zostera*); and from a sketch with which that gentleman has favoured me, and the testimony of the fishermen, it would appear that it has there attained nearly the size of those taken in the Mediterranean; whilst those found on the coast of Cornwall have not exceeded two inches and a quarter.

From the authorities already quoted, we learn that this species abounds amongst the rocks near the coast, in company with various Palemonidæ; and Roux informs us that it is commonly eaten fried, with such smaller Macroura. Its habits are wholly nocturnal, as it hides itself always during the day. Its eggs are deposited in March and August.

The colours of this species are described as very pleasing. The general tint is a yellowish brown; the pincers white, with a slight hue of rose. The scales of the antennæ and those of the tail are fringed with long rose-coloured cilia. Two remarkable varieties are mentioned—one of a delicate rose-colour, and the other a deep yellow, slightly varied with brown.

APPENDIX,

CONSISTING OF SPECIES OBTAINED DURING THE PROGRESS OF THE WORK.

DECAPODA.
BRACHYURA.

CANCERIDÆ.

Xantho tuberculata, R. Q. Couch, m. s.

Specific Character.—Carapace slightly depressed anteriorly; latero-anterior margin with four triangular teeth; hands and wrists tuberculated, rugose; fingers nearly black, the moveable one with three grooves; third joint of the ambulatory legs denticulated on the upper edge.

The carapace is slightly depressed anteriorly, more so than in *X. rivulosa,* but somewhat less than in *X. florida.* The rostrum slightly waved, minutely emarginate; the anterior portion of the carapace somewhat rugose; the regional lines of demarcation sharp and distinct, the elevations slight and flattened; the latero-anterior margin with four triangular teeth. The anterior pair of feet robust, nearly equal; the hands and wrists somewhat transversely tuberculated and rugose; the wrist with two distinct tubercles anteriorly; the moveable finger has three grooves, one on the inner and two on the outer side. The whole of the ambulatory feet have the fourth, fifth, and sixth joints hairy, with longer cilia on the edges; the third joint distinctly denticulated along the upper margin, with a hairy patch beneath.

The general colour of " the carapace is light flesh colour brown; and the first pair of claws almost transparent yellow." The fingers are black or very dark brown.

I have no hesitation in adopting the view of my friend, Mr. R. Q. Couch, in considering this species as distinct, not only from either of those already described in this work, but from all others previously known. It differs from both the former conspicuously in the distinctly tuberculated hands and wrists, in the entire hairy covering of the three terminal joints of the ambulatory feet, and in the denticulated upper margin of their third joint. It differs from *X. rivulosa* and agrees with *X. florida* in the depressed form of the rostrum,—while it agrees with the former and differs from the latter species in the grooving of the moveable finger.

For the discovery of this interesting addition to our British Carcinology, we are indebted to Mr. Richard Q. Couch, of Penzance, who has kindly sent me the only specimen I have seen. He informs me that it appears to prefer deeper water than the other two species, as he found it repeatedly in the crevices of the *Eschara foliacea*, in the deep water off the Runnell Stone, in Mount's Bay. In the summer it approaches the shore and is found under stones. It spawns in June.

The name of *tuberculata* has been given to the species by its discoverer, from whom and from his father, Mr. Jonathan Couch, of Polperro, I have had so many claims upon my acknowledgments for their intelligent and ready assistance in the progress of the present work.

DECAPODA.   *PORTUNIDÆ.*
BRACHYURA.

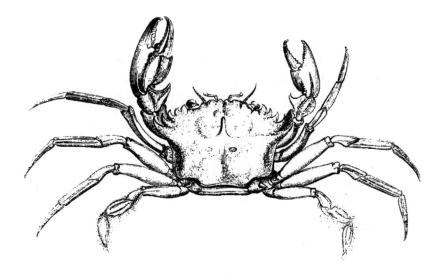

LONG-LEGGED SWIMMING CRAB.

Portunus longipes. Risso.

Specific Character.—Front slightly four-lobed; latero-anterior margin much shorter than the latero-posterior; legs remarkably long.

Portunus longipes,		Risso, Crust. des Env. de Nice, t. i. f. 5, p. 30; Hist. Nat. de l'Eur. Merid. V. p. 4,—Latr. Encycl. X. p. 192.—Roux, Crust. de la Medit. t. iv.—Edw. Hist. Nat. des Crust. I. p. 445.
,,	*infractus,*	Otto, Mem. de l'Acad. de Bonn, XIV. t. xx. f. 1.
,,	*Dalyelii*	S. Bate Annals of Nat. Hist. 1851, p. 320, t. xi. f. 9.

The general form of this interesting species is very different from that of the others of the genus. It is altogether more slight and slender in its proportions. The carapace is flattened, and, in the male, broader than it is long, in the proportion of three to two; in the female the disproportion is not so great. It is divided transversely by a ridge, which terminates at each side in a long and sharp tooth, the posterior of the five which occupy the latero-anterior margin; of these teeth the middle one is broader than the others, and the posterior is much longer; they are all somewhat curved forwards. The anterior portion of the carapace is minutely granulated, and has several slight elevations; the front is slightly four-lobed, the division more strongly marked in the male. The first pair of legs strong and angular; the wrist having a strong tooth on the inner anterior angle; the hand with two carinæ above, the inner one terminating in a small spine. The moveable finger with three distinct longitudinal carinæ, and deep intermediate grooves. The three following pairs of feet long and slender, increasing in length to the fourth, which is the longest of all; flattened, the last three joints longitudinally grooved: the fifth pair slender and weak, the terminal joint lanceolate and slightly grooved. The abdomen, in the male, triangular, the last joint abruptly narrowed; in the female, broad and much rounded.

The colour of this species is a rich deep brownish red, with reddish grey spots; the abdomen yellowish or pinkish white.

The occurrence of this truly Mediterranean species on our southern coast is interesting, as affording another instance of the partial identity of the Fauna of the two shores, to which I have already had occasion so repeatedly to refer. It had not, I believe, been found on our shores until it was dredged on the coast of Cornwall in the year 1848, by my friends Prof. E. Forbes and Mr. M'Andrew, from whom I received a male specimen, and subsequently, through the kindness of Mr. Cocks of Plymouth, a female, which was taken by that gentleman. I also received a specimen from Mr. R. Q. Couch, of Penzance, during the year above-mentioned. It is doubtless the species described by Mr. Spence Bate as new, in the "Annals of Natural History," for 1851, under the name of *Portunus Dalyellii*, from a specimen obtained in Oxwich Bay, near Swansea. The lateral spines are very largely developed in the figure given by Mr. Bate, but not more so than in many Mediterranean specimens, and scarcely more than in Roux's figure. It is at a glance distinguished from all other species by the character from which the name has been given, namely, the length and slenderness of the legs.

DECAPODA.
BRACHYURA.

CORYSTIDÆ.

GENUS THIA, Leach.

CANCER.   Herbst.
THIA.   Leach, Risso, Latr. Edw.

Generic character.—*External antennæ* of moderate length, inserted beneath the front, just at the inner side of the orbits. *Internal antennæ,* transversely folded beneath the front. *External pedipalps* extending forwards to the antennal fossa; the inner branch with second joint shorter than it is broad, its anterior inner angle truncate and submarginate. *Orbits* extremely small. *Eyes* scarcely visible. *Carapace* somewhat heart-shaped, considerably narrowed at the posterior portion, nearly horizontal from before backwards, much arched from side to side; the front broad, lamelliform, entire, bounded by a small notch on each side. *Anterior legs* short, and slightly compressed, the fingers deflexed; the remaining pairs still shorter, each terminating in an acute styliform joint. *Abdomen* in both sexes very narrow; in the male with the three middle joints united; in the female the seven joints all moveable.

A GENUS established by Leach, in his " Zoological Miscellany," on a species of which he was ignorant of the locality.

DECAPODA.
BRACHYURA.

CORYSTIDÆ

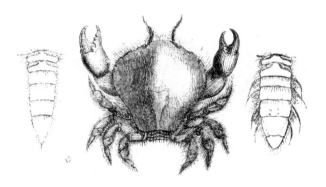

Thia polita, Leach.

Cancer residuus ?	HERBST, t. xlviii. f. 1.
Thia polita,	LEACH, Zool. Miscell. II. t. ciii.—GUERIN, Icon. du règne anim. t. iii. f. 3.—EDW. Hist. des Crust. II. p. 144.
Thia Blainvillii,	RISSO, Hist. Nat. de l'Eur. Merid. v. p. 19.

THE carapace in *Thia polita* is perfectly smooth and polished; the circumference is almost entire, excepting a small notch over each orbit; its outline is contracted towards the posterior portion, and the posterior margin is truncated in the male, and slightly hollowed in the female; it is nearly horizontal from the front to the posterior margin, and much arched from side to side; the front is prominent and evenly arched; the whole of the margin is ciliated with long hair. The orbits are very small, and the minute eyes are ordinarily concealed within them. The anterior pair of legs are robust, the surface polished; the hand is rounded, the posterior outer angle obliquely cut away for the articulation of the wrist; the fingers are slightly deflexed, and armed with a few small tubercles. The remaining legs are shorter than the former, the hinder ones being the shortest; the joints, particularly the penultimate, rounded and somewhat gibbous; the terminal one subulate and acutely pointed, the whole of them strongly ciliated. The abdomen in the male is five-jointed, from the soldering of the middle three joints, but their distinction is still obvious from the transverse groove not being obliterated; it is narrow triangular, the last joint very small: in the female the abdomen is seven-jointed, and a little broader than in the male; in both it is fringed with long hair.

Length, 0·6 of an inch, breadth rather more.

We owe our knowledge of this rare species, as indigenous to Britain, to the researches of Dr. Melville, the learned Professor of Natural History in Queen's College, Galway. It was found by him buried in the sand, and three specimens, one male and two females, obligingly forwarded to me. Both the females were loaded with spawn. Hitherto this is the only instance of its occurrence as a native of our coasts.

The species was first figured by Herbst in his great work, but somewhat imperfectly; I cannot, however, join with Dr. Leach and Milne Edwards in doubting that Herbst's species is identical with that of the individual on which Leach founded his Genus *Thia*. The habitat of that specimen was unknown; but Risso has described, under the name of *Th. Blainvillii*, what I cannot but believe to be this species. It is stated by Dr. Milne Edwards to inhabit "La Manche" and the Mediterranean.

It is right to mention that there is one character in which the Irish specimen differs from the descriptions given by Leach and Milne Edwards; namely, in the much shorter length of the antennæ; but Guerin's figure, in the "Iconographie du règne animal," exactly agrees in this respect with the former.

GENUS DROMIA, Edw.

| CANCER. | Linn. Herbst. |
| DROMIA. | Fabr. Latr. Leach, Edw. |

Generic character.—*External antennæ* placed beneath the ocular peduncle; the auditory tubercle, occupying the base, very large, and perforated at the external angle; the next joint large and nearly cylindrical, forming the inferior boundary of the orbit, and armed with a strong tooth. *Internal antennæ* with the basal joint nearly cylindrical; the antennary fossæ longitudinal and distant, and incomplete at the outer side. *Anterior feet* very robust, terminating in a strong claw, the extremities strongly toothed, and spoon-shaped. The *second and third pairs of feet* of moderate and nearly equal length, terminating in a sharp somewhat curved nail; the basal joint of the third in the female pierced with the opening of the generative apparatus; the *fourth and fifth pairs* very small, turned over the back of the carapace, against which they are closely pressed, each terminating in a small but perfect double claw. *Carapace* somewhat globular, the regions distinctly marked; the front inclined and small. *Orbits* deep. *Eyes* with short peduncles.

The characters of this remarkable genus are, to a certain extent, intermediate between the brachyurous and macrourous forms. In the young state the great predominance of the posterior or abdominal regions of the body approximate it in some measure to the latter; and the general form of the cephalo-thoracic portion, especially in the adult condition, is not less assimilated to the former division of the class. The characters of the posterior pairs of feet at once remove it from either of these, and indicate its true place to be amongst those anomalous forms which have been associated by Dr. Milne Edwards into the intermediate group, the ANOMOURA.

The species of this genus are very widely distributed. The Indian and African shores, those of the Red Sea and of the Mediterranean, the islands of the West Indies, and the coasts of South America, have furnished various species; and our own southern coast has of late years been found to give a place of habitation to one of the most conspicuous species.

DECAPODA.   *DROMIADÆ.*
ANOMOURA.

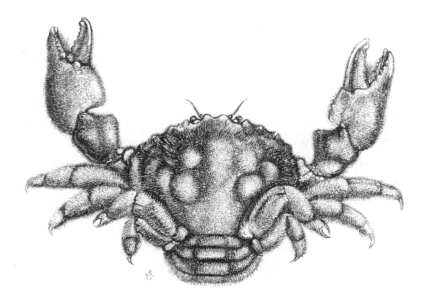

Dromia vulgaris, Edwards.

Specific Character.—Carapace broader than it is long; latero-anterior margin with four strong teeth, the second having a tubercle at its base; the last joint of the abdomen in the male broader than it is long.

Cancer dromia,   " Olivi Zool. Adriat. p. 45" (M. EDW.).
Dromia Rumphii,   BOSC, Hist. des Crust. I. p. 229.—DESMAR. Consider. sur les Crust. p. 137.—BLAINV. Fauna Franc. Crust. t. vii. fig. 1.—RISSO, Hist. Nat. de l'Eur. Merid. V. p. 32.—EDW. Règ. anim. de Cuv. Edit. 3. Crust. II. f. 1.
Dromia vulgaris,   EDW. Hist. Nat. des Crust. II. p. 173.

THIS species, which has at length been undoubtedly proved to inhabit our southern shores, and probably too in considerable numbers, has the carapace strongly knobbed above, especially at the anterior portion, and very much raised so as to approach the globular form; the front with three teeth, which become less prominent by age; there is a fissure above the external angle of the orbit, and a tooth beneath that cavity. The latero-anterior margin has four strong teeth, the bases of which are long; the first situated beneath the line of the orbit, the second furnished near its base with a tubercle or small secondary tooth, thus appearing almost as if double; the third occupying a larger portion of the margin than either of the others, and the last the smallest. The latero-posterior margin nearly as long as the latero-anterior. The first pair of legs are robust and nodulated; the hand has several small conical teeth on its upper and inner edge. The claws are smooth and polished, strongly denticulated and internally hollowed at the extremity, the denticles of each finger shutting into the interspaces of the other; the moveable finger much curved on the upper side; the wrist largely nodulated; the second and third pairs of legs much shorter than the first, terminating in a strong, sharp, curved nail; the fourth and fifth pairs are doubled back over the posterior part of the carapace, flattened, and each terminating in a sharp, tolerably perfect, double claw. The abdomen in the male is much curved longitudinally, and the joints are distinct; the terminal one broader than it is long. In the female the abdomen is extremely broad and much curved; each joint elevated in the centre, and on each side. The whole animal, body and limbs, covered with dense short hair, which in the young state is of a buff colour, and in the adult dull brown.

Length of the carapace of a full grown male two inches and a half, breadth three inches.

I have carefully examined the hair with the microscope, in individuals of various ages, and have not found in any one instance the club-shaped hair assigned by Dr. Milne Edwards to this species. The hair is in all cases setaceous, very acute at the point, and is itself furnished with minute hairs along its sides.

The first intimation of the present species as a native of Britain, occurs in an announcement by Mr. John Edward Gray, at a meeting of the Zoological Club of the Linnæan Society, as long since as June 22nd, 1824. These were stated to have been seen by that gentleman in Billingsgate Market, amongst some oysters, which had been brought from Whitstable Bay, on the Kentish coast. This fact is recorded in the " Zoological Journal," Vol. I. p. 419. In the " Zoologist," 1848, p. 2325, occurs a notice of no fewer than nine full sized specimens having been dredged on the coast of Sussex. Mr. Newman gives the details of its occurrence, and a figure of the species, having received it from Mr. George Ingall. About the same time my lamented friend Mr. Dixon, of Worthing, sent me three specimens which had been procured off Selsey Bill. Mr. Newman alluding to Linnæus's name of an allied species, *cancer* " *dormia*," supposes it to refer to its sedentary and lethargic habits. Linnæus was, however, too good a scholar thus to render a derivative from *dormio*; it is plainly a misprint for *dromia*, from the Greek $\Delta\rho\acute{o}\mu\omega\nu$, a little running crab; and in the " Amœnitates Academicæ," Linnæus himself gives the correct spelling.

I some years since received numerous young specimens from Sicily, every one of which had the carapace entirely covered with a sponge which had grown over it, concealing even the two hinder pairs of legs, which were closely pressed against the back, and rendered immoveable. It is a common Mediterranean species.

DECAPODA.
ANOMOURA.

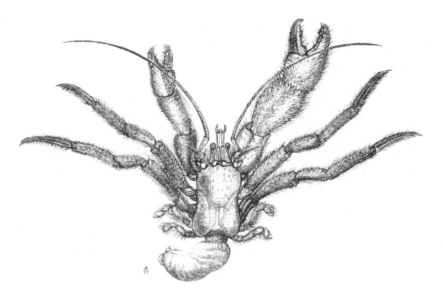

Pagurus Thompsoni, mihi.

Specific Character.—The whole of the legs hispid and spinous; anterior pair unequal; the wrist as long as the hand; eye stalks extending to half the length of the last joint of the peduncle of the external antennæ; antennal spine curved outwards, and furnished with a row of small spines on the outer edge.

The carapace is polished, but sparsely marked with impressed dots; the front nearly entire. The eye-stalks are cylindrical, and furnished with a regular longitudinal series of minute tufts of hair along the upper surface; they extend forwards to the middle of the last joint of the peduncle of the external antennæ. The antennal spine is curved outwards, spinous and hairy on its outer edge. The internal antennæ are half as long again as the peduncle of the external. The anterior feet very unequal, bristly, and spinous; the larger hand twice as long as it is broad, hairy, beset with spinous tubercles, of which there is a stronger series along the outer side; the moveable finger with a strong tubercle fitting between two smaller ones on the other finger when closed; wrist about as long as the hand, and equally hairy and spinous, with a row of longer spines along the inner edge. Smaller anterior leg nearly linear, the proportions, clothing, and armature somewhat similar to the larger, but the opposing edges of the fingers without tubercles. Third and fourth pairs of feet very long, covered with stiff hairs and small spines; the last joint armed with a series of strong spines along the inferior edge, and terminated by a sharp nail.

The general aspect of this species reminds one of *P. Prideauxii*, the proportions of the parts being somewhat similar; but it differs not only in some proportional characters, but strikingly in the spinous and hispid clothing of the whole of the legs. It bears in these latter circumstances some relation to *P. Cuanensis*, but from this it may be distinguished by the proportions between the wrist and hand, the form of the wrist, the relative proportions of the eye stalks and antennal peduncle, and other characters.

I have a melancholy gratification in dedicating this species by name to a gentleman who for many years was justly considered as the representative of the Zoology of Ireland, and whose acute discrimination and persevering enthusiasm in his favourite pursuit, were only equalled by the liberal and unselfish feeling with which he placed his treasures in the hands of his fellow labourers, whenever he believed the interests of science would be thereby furthered. The specimen from which the above description is taken, was placed in my hands by my lamented friend only a very few days before his untimely death deprived the science of Ireland of one of its most distinguished ornaments, and society of as kind and true hearted a man as ever lived.

Mr. Thompson's note given me with the specimen is as

follows:—" Dredged at fifty fathoms, entrance of Belfast Bay, by Mr. Hyndman." It was in the shell of the common whelk, *Buccinum undatum*.

The vignette is from a tesselated Roman pavement discovered at Cirencester in 1783.

DECAPODA. *PAGURIDÆ.*
ANOMOURA.

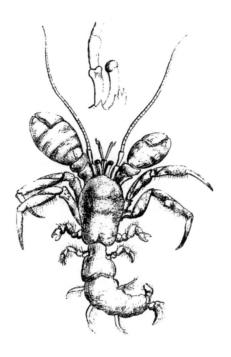

BLUE-BANDED HERMIT CRAB.

Pagurus fasciatus. Mihi.

Specific Character.—Anterior legs unequal; hand oval, smooth; eye-stalks as long as the penultimate joint of the external, and nearly half as long as the whole of the internal antennæ; body and legs banded alternately with red and blue.

The carapace is smooth and even, ovate, rounded in front, truncated and slightly emarginate behind. External antennæ as long as the whole of the body, the peduncle cylindrical, the second and last joints of nearly equal length, and apparently without any spine. The internal antennæ are of moderate length, less than twice as long as the peduncle of the external. Eye stalks nearly cylindrical, slightly curved outwards, as long as the penultimate joint of the external antennæ. The anterior pair of feet are robust, of unequal size; the hand is oval, broader anteriorly, slightly pointed at the extremity of the fingers; the moveable finger fitting the other closely; the wrist subquadrate, broader than it is long; the second and third pairs with the penultimate joint ciliated on the inner edge.

The body is obscurely, and the whole of the legs distinctly marked with alternate bands of red and blue.

The whole of the above description is given from a coloured drawing, for which I am indebted to Mr. Cocks, of Falmouth, and from which also the woodcut is taken. It was obtained by him at Falmouth. I have never seen a specimen, but I am confident that Mr. Cocks's accuracy of delineation may be implicitly relied on.

This species may at once be distinguished from every other known on our coasts. The only one to which, from the form of the hands, it bears a *primâ-facie* resemblance, is *P. Hyndmanni*, but from this it differs in the form of the thorax, the comparative length of the internal antennæ, and many less obvious characters. The distinct alternate bands of blue and red render it one of the most beautiful of the genus.

DECAPODA.   PAGURIDÆ.
ANOMOURA.

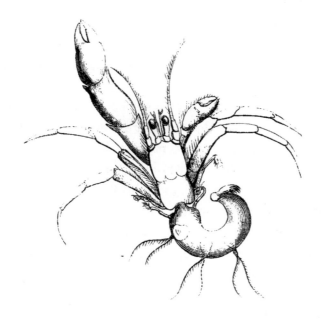

"*Pagurus Dilwynii.*"   Sp. Bate.*

"CARAPACE smooth and polished. Colour bluish, marked with brown.

"First pair of feet unequal, the *left* being much longer than the right; smooth to the naked eye, but under a lens perceived to be minutely granulated. The second and third joints are armed with teeth, which give the limb an angular character. The *right* is very short and covered with hair.

"The external antenna is about two-thirds the length of the longest of the first pair of feet, and hairy; its base as long as the eye-stalks, which are slender and long. The basal tooth, with which the antenna of this genus is generally armed, is wanting.

* Annals of Nat. Hist. 1851, p. 320, pl. x. fig. 11.

"The false feet in the female are *long* and feathery, and divide at the *base*.

"The most striking difference between this and other British species of the *Pagurida*, is exhibited in the form of the first pair of feet, and the length of the external antennæ.

"Having met with only this solitary specimen, it is impossible to say but that the right foot of the first pair, which is usually the longer, may be in the process of being reproduced from loss; although I am inclined, from its well-developed character, to believe that the left is in this species the more important of the two. The false feet, which in the female are generally forked, are so in this specimen, but very much nearer to the base than in the common species.

"It burrows very rapidly in the sand. Taken near the Worms Head, Swansea.

"Mr. Couch has informed me, since this has been in the hands of the printer, that he has also found the species in Cornwall.

"The name applied to this species is one long-known to science, and honoured as the stimulator of natural history in this locality in the person of L. W. Dillwyn, Esq., Sketty Hall."

The foregoing description and figure are copied from those given by Mr. Spence Bate, in the "Annals of Natural History," as I have never seen the species.

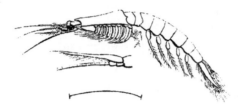

GENUS CYNTHIA. J. V. Thompson.

Generic character.—" Subabdominal fins composed of two joints, four last fins with the terminal plume double, with an opake bifurcate, and convolute organ rising between each."

Cynthia Flemingii.   H. Goodsir.

Specific Character.—" Inferior antennal scale almost twice as long as the peduncle. A thick fringe of strong hairs bordering its edge. Rostrum slender and finely pointed. Volute organ between the plumose setæ of the subabdominal fins minute ; edges of the middle plate of the tail spined.

" Long. eight lines.   Hab. Frith of Forth.

" *Description.*—The whole body of an opake straw-colour, with the reticulated portions of the eyes black. Superior antennæ with the peduncle three-jointed, the two cetaceous portions arising from the second joint of the peduncle, the last joint ovate, surrounded with a thick fringe of hairs ; these hairs are bent downwards at their extremities, so as to form a concavity on the lower surface. The peduncle is about twice the length of the eyes. The peduncle of the inferior antennæ extends to the origin of the setaceous portion of the superior antennæ ; the two last joints are slender and clavate. A long, slender, and pointed scale arises from the first joint of the peduncle, above the setaceous portion; this is twice as long as the peduncle, and is thickly fringed with long hairs, which are directed inwardly so as to meet those of the opposite side. The carapace is not very large, curved at its posterior edge, and produced at its posterior and inferior angle.

"Abdomen slender, the inferior edge of each segment considerably produced, and all of them but the last bearing a fin composed of two joints; the first joint is scale-like clavate; the second is multiarticulate and plumose; all of them but the first pair double. The bifurcate convolute organ, between the double plumes, is very minute. Middle plate of the tail edged with spines on its sides, and entire at the extremity. External caudal fins twice as long as the middle plate, and pointed.

"The bifurcate and convolute organ between the double plumes of the four last subabdominal fins, together with the number of joints in these fins, seem to be the most striking characters of this genus. Mr. Thompson, in the third memoir of his 'Zoological Researches,' says, 'It is not in the number of joints alone, however, that they (subabdominal fins) differ, their form and structure is also essentially different. In *Cynthia* the four last of these members are each composed of a very large bilobate scale, supporting at its apex two taper articulate fins, strongly ciliated with plumose setæ; from between these originates an opake organ which bifurcates, its two extremes of unequal length being rolled inwards, the one over the other.'

"Mr. Edwards considers that these last are the branchial apparatus.

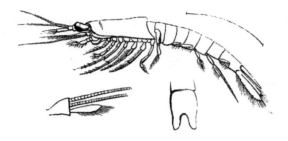

GENUS THEMISTO. H. Goodsir.

Generic character.—" External antennæ armed with a scale. First, second, and fifth segments of the abdomen bearing fins like the mysis. Third and fourth with the peduncles bi-articulate, and each peduncle giving off two branches; the external branch of the fourth very long and slender, semi-articulated.

Themisto longispinosa. H. Goodsir.

Specific Character.—" Superior antennal scale of the same length as the terminal joint of its peduncle; armed at its extremity with a thick tuft of hairs. Inferior antennal scale twice as long as its peduncle; fringe not strong. Third subabdominal fin with its internal branch minute. Internal branch of the fourth with a few long hairs from the extremity only. External branch reaching from the extremity of the caudal fins. Internal caudal fin truncated.

"Long. three-quarters of an inch. Hab. Frith of Forth.

" *Description.*—The whole body of a dark yellowish or greenish colour. Eyes large, reaching to the extremity of the peduncle of the inferior antennæ. The reticulated portion black, and produced backwards inferiorly. Rostrum very short but sharply pointed. First joint of the peduncle of the inferior antennæ very strong, the two following slender; the setaceous portion of the antennæ arising from the extremity of the last. The scale arises from the inner and superior part of the first joint of the peduncle; it is hardly twice the length of the peduncle, slender, and tapering very gradually to the extremity; it is rather thinly fringed. The upper surface of the peduncles of the superior antennæ hollowed out, forming a bed for the eyes. A short ovate scale arises from the inferior part of the last joint, immediately below the origins of the setaceous portions of the antennæ. A thick bunch of matted hair arises from its extremity, which gives it the appearance of being bi-articulated. The inferior edge of the external seta of the superior antennæ bears a thin fringe of very strong hairs, which are thickest and strongest near the base. The carapace is not large, leaving two of the thoracic segments exposed posteriorly; it is rounded at its anterior and inferior angle, and considerably produced at its inferior and posterior angle. A strong bi-articulate and chelate palpus arises from each side of the mouth. The abdomen is slender, but the segments are not produced inferiorly. The branchial subabdominal fins are five in number; they arise from all the abdominal segments except the last [two]. The first, second, and fifth are like those in the genus *Mysis*, namely, a single plumose joint; the third and fourth are pedunculated,—the peduncles being composed of two joints. The first joint is minute, the second is of considerable length; two branches arise from the extremity of the second joint; these branches, in the third fin, are both plumose; in the fourth one the internal only is plumose. The external branch of the fourth consists of a very long six-jointed spine, which reaches beyond the extremity of the caudal fins; it is very finely pointed; the internal branch about the same as the first joint of the external branch. The caudal plate is slightly swollen near the base; its edges are serrated, and its extremity bifurcated; the bottom of the fourth being rounded, and the extremities of the fork also blunted and rounded. The internal caudal fins are truncated at their extremities; the external are paddle-shaped, and rounded at their extremities. Both of these fins are fringed at their extremities and inferior edges with long hairs."

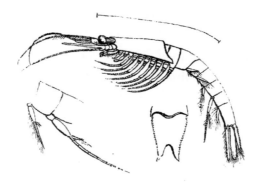

Themisto brevispinosa. H. Goodsir.

Specific Character.—" Superior antennal scale not so long as the peduncle. Inferior antennal scale four or five times as long as the peduncle. Internal branch of the third subabdominal fin minute; the internal branch of the fourth longer than the first joint of the external branch; the external branch extending a little beyond the base of the caudal fins, ending by means of a dart-like point. The lateral caudal fin ending in a sharp point superiorly, and rounded inferiorly; the internal fin oblong, ovate, and pointed. The lateral edges of the middle plate bearing a single row of long, sharp, and bent spines, contracted near the base and the bottom of the fork, forming an acute angle; prongs pointed.

" Long. one inch. Hab. Frith of Forth.

" *Description.*—The whole body more robust than that of the last-described species, and of an opake white colour, with a single row of black spots along the dorsal mesial line of the abdominal segments. The first joint of the peduncle of the inferior antennæ very short and almost circular; the two following are slender. The scale which arises from the superior part of the first joint above the true antennæ is very strong at the base, and then tapers gradually to a fine point. A fringe of long hairs borders its inferior edge. These hairs are matted at the extremity so as to give them the appearance of a second joint; two or three short strong spines arise from the extremity of the scale. The third joint of the peduncle of the superior antennæ is considerably produced at its superior angle. The scale which arises beneath the setaceous portions is strong, bent upwards at its extremity, and pointed, but not fringed. The eyes are large; the reticulated portion circular.

"The rostrum is of considerable length, but it is not sharp. The internal branch of the third subabdominal fin is minute; the external one is long, slender, and finely pointed; it is also fringed with very long hairs. The internal branch of the fourth fin is longer than the first joint of the external branch; and it is both more strongly ringed and more moveable than that of the last-described species. The external branch extends a little beyond the base of the caudal fins. The sixth or last joint of this branch suddenly contracts near the extremity to about half its original thickness, ending in a dart-like point. The external caudal fins end in a sharp point inferiorly, and are rounded inferiorly; the internal fins are oblong, oval, and pointed at the extremity. These are both fringed at their inferior edges and at their extremities. The lateral edges of the middle plate armed with a single row of strong hooked spines. It is contracted near the base, and the angle formed by its bifurcation is very acute; the extremities of the prongs are also sharp-pointed, and of a black colour."

Amongst the numerous and interesting additions to our knowledge of the smaller Crustacea, for which we are indebted to Mr. H. D. Goodsir, are the three foregoing species of the family MYSIDÆ. As I have never seen specimens of either of them, I have thought it best to give, verbatim, Mr. Goodsir's own descriptions, although somewhat diffuse, with copies of his figures.

In addition to the species included in this Appendix, I may refer to some which have been described in recent periodical publications, from the observations of Mr. Spence Bate, Mr. William Thompson of Weymouth, and others. I have not transferred these descriptions to this work, partly because I am not in all cases quite satisfied with the grounds on which the species have been considered as distinct, and because they may be readily examined in their original place of publication. At the same time I am anxious to express my gratification at this accession of young intelligent naturalists in this field of observation, from whose active and continued labours the most valuable results may be anticipated.

Woodfall and Kinder, Printers, Angel Court, Skinner Street, London.

PREPARING FOR PUBLICATION,

A HISTORY

OF THE

BRITISH SESSILE-EYED CRUSTACEA.

By J. O. WESTWOOD, F.L.S., ETC.

About Six Parts will complete the work; these will be published monthly, at 2s. 6d. each. Part I. will be published shortly. The mode of illustration will be similar to that adopted in Professor Bell's History of the Stalk-eyed species.

Mr. Westwood solicits notices of habitats, &c., from those who have made this interesting portion of the Crustacea their particular study.

THE NATURAL HISTORY

OF

THE BRITISH ISLES.

This Series of Works is Illustrated by many Hundred Engravings; every Species has been Drawn and Engraved under the immediate inspection of the Authors; the best Artists have been employed, and no care or expense has been spared. A few copies have been printed on larger Paper.

QUADRUPEDS, by PROFESSOR BELL. A New Edition preparing.

BIRDS, by MR. YARRELL. Second Edit., 3 vols. 4*l*. 14*s*. 6*d*.

COLOURED ILLUSTRATIONS OF THE EGGS OF BIRDS, by MR. HEWITSON. A New Edition, 6 Parts published, at 2*s*. 6*d*. each.

REPTILES, by PROFESSOR BELL. Second Edition, 12*s*.

FISHES, by MR. YARRELL. Second Edition, 2 vols. 3*l*.*

CRUSTACEA, by PROFESSOR BELL.

STAR-FISHES, by PROFESSOR EDWARD FORBES. 15*s*.

ZOOPHYTES, by DR. JOHNSTON. Second Edition, 2 vols. 2*l*. 2*s*.

MOLLUSCOUS ANIMALS AND THEIR SHELLS, by PROFESSOR ED. FORBES and MR. HANLEY. 4 vols. 6*l*. 10*s*.; or Large Paper, with the Plates Coloured, 13*l*.

FOREST TREES, by MR. SELBY. 28*s*.

FERNS, by MR. NEWMAN. Third Edition. Now in the Press.

FOSSIL MAMMALS AND BIRDS, by PROFESSOR OWEN. 1*l*. 11*s*. 6*d*.

A GENERAL OUTLINE OF THE ANIMAL KINGDOM, by PROFESSOR T. RYMER JONES. 8vo. A New Edition preparing.

* "This book ought to be largely circulated, not only on account of its scientific merits—though these, as we have in part shown, are great and signal—but because it is popularly written throughout, and therefore likely to excite general attention to a subject which ought to be held as one of primary importance. Every one is interested about fishes—the political economist, the epicure, the merchant, the man of science, the angler, the poor, the rich. We hail the appearance of this book as the dawn of a new era in the Natural History of England."—*Quarterly Review*, No. 116.

JOHN VAN VOORST, 1, PATERNOSTER ROW.

Preparing for publication, a New Edition, with Illustrations, of
THE NATURAL HISTORY OF SELBORNE,
By the late Rev. GILBERT WHITE, M.A. With a brief Memoir of the Author, and extracts from his Diary and Correspondence. Edited by THOMAS BELL, F.R.S., President of the Linnean Society, Professor of Zoology in King's College, London.

JOHN VAN VOORST, 1, Paternoster Row.

OTHER BOOKS PUBLISHED BY MR. VAN VOORST.

ANSTED, Professor. AN ELEMENTARY COURSE OF GEOLOGY, MINERALOGY, AND PHYSICAL GEOGRAPHY. Post 8vo. Illustrated, 12s.

THE ANCIENT WORLD, with 149 Illus. A New Edition, Post 8vo, 10s. 6d.

BABINGTON, CHARLES, C., M.A. A MANUAL OF BRITISH BOTANY. Third Edition, 12mo, 10s. 6d.

BAPTISMAL FONTS. A Series of 125 Engravings with Descriptions. 8vo, 1l. 1s.

BURTON'S (R. F.) FALCONRY IN THE VALLEY OF THE INDUS. Post 8vo, 4 Illustrations, 6s.

COUCH, JONATHAN, F.L.S., ILLUSTRATIONS OF INSTINCT, deduced from the Habits of British Animals. Post 8vo, 8s. 6d.

DALYELL, SIR JOHN GRAHAM, BART. RARE AND REMARKABLE ANIMALS OF SCOTLAND, Represented from Living Subjects: with Practical Observations on their Nature. 2 vols. 4to, containing 109 Coloured Plates, 6l. 6s.

THE POWERS OF THE CREATOR DISPLAYED IN THE CREATION: Or, Observations on Life amidst the various forms of the Humbler Tribes of Animated Nature. In 2 vols. 4to, 115 Plates, 4l. 4s. each.

DOWDEN'S (RICH.) WALKS AFTER WILD FLOWERS. Fcap. 8vo, 4s. 6d.

DRUMMOND, Professor. FIRST STEPS TO ANATOMY. Plates, 12mo, 5s.

ENGLAND BEFORE THE NORMAN CONQUEST. 16mo, cloth, 2s. 6d.

ELEMENTS OF PRACTICAL KNOWLEDGE. Second Edit. 16mo. Illustrated, 3s.

EVENING THOUGHTS. By a Physician. Post 8vo, 4s. 6d. Second Edition.

GOLDSMITH'S VICAR OF WAKEFIELD. With 32 Illustrations by MULREADY, 1l. 1s., square 8vo, or 36s. in morocco.

GRAY'S ELEGY IN A COUNTRY CHURCHYARD. With 33 Engravings. Post 8vo. 9s.—A Polyglot Edition, with inter-paged Translations in the Greek, Latin, German, Italian, and French languages. 12s.

GRAY'S BARD. With Illustrations. Uniform with the Elegy, 7s.

HARDING, W. UNIVERSAL STENOGRAPHY. 12mo, 3s. sewed. 3s. 6d. bound.

HARVEY, Professor, THE SEA-SIDE BOOK: the Natural History of the British Coasts. Second Edition. Fcap. 8vo, with 69 Illustrations, 5s.

A MANUAL OF THE BRITISH MARINE ALGÆ (Sea-Weeds), with Plates of the Genera. 8vo, 21s.; coloured copies, 31s. 6d.

NEREIS BOREALI-AMERICANA. Royal 4to, Part I., with coloured plates, 15s.

HENFREY, ARTHUR, F.R.S. THE RUDIMENTS OF BOTANY. 16mo, with Illustrative Woodcuts, 3s. 6d.

OUTLINES OF STRUCTURAL AND PHYSIOLOGICAL BOTANY. With 18 plates, Foolscap 8vo, 10s. 6d.

VEGETATION OF EUROPE: its Conditions and Causes. Foolscap 8vo, 5s.

HEWITSON, WILLIAM C., F.L.S. EXOTIC BUTTERFLIES: Illustrations and Descriptions of New Species (to be continued Quarterly), 5s.

INSTRUMENTA ECCLESIASTICA: a Series of designs for the Furniture, etc., of churches and their precincts. 4to, 1l. 11s. 6d.—A second series is now in course of publication in parts at 2s. 6d. each.

JENYNS, THE REV. LEONARD. OBSERVATIONS IN NATURAL HISTORY. Post 8vo, 10s. 6d.

JESSE, EDWARD, F.L.S. AN ANGLER'S RAMBLES. Post 8vo, 10s. 6d.

JONES, Professor T. RYMER. THE NATURAL HISTORY OF ANIMALS. Vol. I. with 105, and Vol. II. with 104, Illustrations, post 8vo, each 12s.

KNOX, A. E., M.A. GAME BIRDS AND WILD FOWL: their Friends and their Foes. With Illustrations by Wolf. Post 8vo, 9s.

KNOX, R., M.D., F.R.S.E. GREAT ARTISTS AND GREAT ANATOMISTS; a Biographical and Philosophical Study. Post 8vo. 6s. 6d.

LATHAM, ROBERT GORDON, M.D., F.R.S. THE NATURAL HISTORY OF THE VARIETIES OF MAN. 8vo, Illustrated, 21s.

ETHNOLOGY OF THE BRITISH ISLANDS, COLONIES, EUROPE, MAN AND HIS MIGRATIONS. Foolscap 8vo, 5s. each.

PALEY, F. A., M.A. A MANUAL OF GOTHIC MOLDINGS. Second Edition. Illustrated. 8vo, 7s. 6d.

A MANUAL OF GOTHIC ARCHITECTURE. Foolscap 8vo. 70 Illus. 6s. 6d.

PRESTWICH, JOSEPH, Jun., F.G.S., &c. A GEOLOGICAL INQUIRY RESPECTING THE WATER-BEARING STRATA OF THE COUNTRY AROUND LONDON. 8vo, a Map and Woodcuts, 8s. 6d.

SHARPE, EDMUND, M.A. A TREATISE ON THE RISE AND PROGRESS OF DECORATED WINDOW TRACERY IN ENGLAND. Illustrated with 97 Woodcuts and 6 Engravings. 8vo, 10s. 6d.

SPRATT, Capt. T., R.N., and FORBES, Professor. TRAVELS IN LYCIA, MILYAS, AND THE CIBYRATIS. With Illustrations. 2 vols. 8vo, 36s.

THE POOR ARTIST; or, Seven Eye-Sights and one Object, Foolscap 8vo, 5s.

WARD, N. B., F.R.S. ON THE GROWTH OF PLANTS IN CLOSELY-GLAZED CASES. A New Edition, illustrated, post 8vo, 5s.

WATTS' DIVINE AND MORAL SONGS. With 30 Illustrations by COPE. Square 8vo, 7s. 6d., or 21s. in morocco.